Lecture Notes in Computer Science   9510

Commenced Publication in 1973
Founding and Former Series Editors:
Gerhard Goos, Juris Hartmanis, and Jan van Leeuwen

More information about this series at http://www.springer.com/series/8637

Abdelkader Hameurlain · Josef Küng
Roland Wagner · Hendrik Decker
Lenka Lhotska · Sebastian Link (Eds.)

Transactions on Large-Scale Data- and Knowledge-Centered Systems XXIV

Special Issue on Database- and Expert-Systems Applications

 Springer

Editors-in-Chief

Abdelkader Hameurlain
IRIT, Paul Sabatier University
Toulouse
France

Josef Küng
FAW, University of Linz
Linz
Austria

Roland Wagner
FAW, University of Linz
Linz
Austria

Guest Editors

Hendrik Decker
Universidad Politécnica de Valencia
Valencia
Spain

Lenka Lhotska
Czech Technical University
Prague
Czech Republic

Sebastian Link
University of Auckland
Auckland
New Zealand

ISSN 0302-9743   ISSN 1611-3349   (electronic)
Lecture Notes in Computer Science
ISBN 978-3-662-49213-0   ISBN 978-3-662-49214-7   (eBook)
DOI 10.1007/978-3-662-49214-7

Library of Congress Control Number: 2015943846

Printed on acid-free paper

This Springer imprint is published by SpringerNature
The registered company is Springer-Verlag GmbH Berlin Heidelberg

Preface

The 25th International Conference on Database and Expert Systems Applications (DEXA 2014), with proceedings published in volumes 8644 and 8645 of Springer's *Lecture Notes in Computer Science*, featured some outstanding keynote presentations and regular articles. As with previous editions of the conference, the program co-chairs of DEXA 2014 invited some of the authors to submit extended papers to a special issue of the Springer journal *Transactions on Large-Scale Data- and Knowledge-Centered Systems* (TLDKS). Following these invitations, one keynote paper and eight regular articles were submitted. Apart from the keynote paper, each submission was carefully assessed by at least two (often more) recognized experts in the respective field. In two rounds of assessment, 35 reviews were received, most of them of very good quality. In the end, six of the eight regular papers were accepted for inclusion in this special issue, in addition to a revised and extended keynote paper.

The contributions in this special issue address a wide range of important contemporary subject areas in data-centric systems and applications, including reflective modeling, big data, similarity search, large-scale data replication, bioinformatic workflows, data pricing, and data anonymization. In good DEXA tradition, all contributions distinguish themselves by the novelty and innovation they bring to these subject areas.

The keynote paper is authored by Dirk Draheim, who, apart from teaching, leads the Information Technology Services Centre at the University of Innsbruck (Austria), and in particular the High-Performance Computing Department of that center. He is a distinguished expert in the field of systems modeling. Theoretical work on modeling usually focuses on logical abstractions of data structures, objects, processes, and object-level language constructs. Additionally, meta-data are of increasing relevance and importance in many fields of information processing. Hence, what is also needed, yet rarely provided, are means to make metadata accessible on the object level. The main contribution of the keynote paper, entitled "Reflective Constraint Writing," with subtitle "A Symbolic Viewpoint of Modeling Languages," consists in a formal treatment of how to add reflection to object-oriented constraint languages. The work presented in this paper is extensive, covering both introspective as well as manipulative data access, for a large variety of purposes, such as to make models robust against unwanted updates, to give precise semantics to existing modeling language constructs, to enable a more adequate system analysis, to assure the quality of system design, and so on.

The article "PPP-Codes for Large-Scale Similarity Searching" is co-authored by David Novak and Pavel Zezula from Masaryk University, Brno, Czech Republic, and supported by the Czech Research Foundations. Their research addresses the challenging problem of efficiently identifying objects in large search spaces that are similar to a given object. The contribution is a two-phase search algorithm on top of a new sophisticated data structure called PPP-Code index. Phase one computes independent rankings based on a given distance function, while phase two aggregates these rankings to access similar objects sooner. Experiments with artificial and real-world data show that the algorithm reduces the size of output candidates by up to two orders of magnitude while preserving the quality of the answer.

Mouhamadou Ba, Sébastien Ferré, and Mireille Ducassé from IRISA/INSA Rennes and the University of Rennes, France, co-authored the article "Solving Data Mismatches in Bioinformatics by Generating Data Converters." Their research addresses the prevalent problem where different bioinformatics services with mismatches between given outputs and required inputs need to be composed into workflows. The main contribution is an automatic converter that utilizes a rule-based convertibility detection mechanism. Experiments with real-world data types and services from the bioinformatics domain yielded new composition strategies that domain experts were not made aware of by existing ad-hoc approaches.

The paper entitled "A Framework for Sampling-Based XML Data Pricing" has been written by Ruiming Tang, Antoine Amarilli, Pierre Senellart, and Stephane Bressan. The authors are affiliated to the National University of Singapore, the National Scientific Research Centre in Paris (France), or to both. The paper presents a sharp-witted approach to determining the market value of XML data based on data samples. The price of the data depends on the degree of completeness of sampling and on the contextual quality of data. In other words, data completeness can be traded for a discount price. The paper is one of the first to reflect the growing perception of data as merchandizing objects. In fact, the importance of data pricing is very likely to increase with the expected expansion of electronic information trading. Hence, for future work in that growing trend, this paper can be expected to become a point of reference.

The authors Nikolaos Nodarakis, Evaggelia Pitoura, Spyros Sioutas, Athanasios Tsakalidis, Dimitrios Tsoumakos, and Giannis Tzimas of the paper "*kdANN+: A Rapid AkNN Classifier for Big Data*" work at the universities of Patras, Ioannina, the Ionian University in Corfu, and the Technological Educational Institute in Patras, Greece. They propose the use of kNN classification for multidimensional objects. Their paper reports on a novel application in the area of big data, based on the all k-nearest neighbor query method. A divide-and-conquer strategy is pursued: Data space decomposition techniques are deployed for reducing the demand of computational resources. The authors have verified the viability of their solution on experimental data sets. By increasing the dimensionality of the dataset, the total execution cost may exceed the computational power of the cluster infrastructure at hand. To cope with that, the authors propose dimensionality reduction techniques. Their results exhibit differences of computation time and cost between the examined algorithms kdANN and kdANN+, with regard to space dimensionality, the granularity of space decomposition, and the number of nearest neighbors.

The paper entitled "Optimizing Inter-Data-Center Large-Scale Database Parallel Replication with Workload-Driven Partitioning" is authored by Zhen Gao, Hong Min, Xiao Li, Jie Huang, Yi Jin, and An Lei. They are affiliated to various IBM labs in the USA or China, to Tongji University in Shanghai (China), or Pivotal Inc. in Beijing (China). The authors propose two algorithms in order to, firstly, partition a large number of workload-associated tables into a minimal number of point-in-time consistency groups that respect some latency constraints, and, secondly, to refine such partitioning by minimizing the number of transaction splits among the resulting consistency groups. Point-in-time consistency is an important property of replicated data and a critical objective of distributed database management systems. In particular, it is relevant to the design of data repositories that need to be able to handle unexpected shut-downs, so that replica consistency can be recovered. Given the growing use of replication for handling critical big data management challenges, the potential impact of this paper is obvious.

The paper entitled "Anonymization of Data Sets with NULL Values" is authored by Margareta Ciglic, Johann Eder, and Christian Koncilia, from the Alpen Adria University at Klagenfurt (Austria). The paper deals with the problem of anonymizing data with missing or unknown values. NULL values usually represent epistemic gaps, and, in the context of this paper, should not be confused with values that have been used deliberately in order to anonymize data. A solution to the problem of anonymizing data sets with unknown component values has been missing. Rather, database table rows containing NULL-valued attributes are offhandedly discarded in conventional approaches. That, however, may easily yield a disturbing loss of information, which may distort data analysis results and thus lead to faulty decision making. Thus, the paper meets an evident desirability of solutions that do not ignore NULL values, in particular in the context of large volumes of data with a big informational and structural variety, where NULL values are ubiquitous.

To conclude, we would like to thank all authors for their contributions to this special issue. Also, we are grateful to all reviewers for their invaluable work in assessing the papers, thus contributing to the high quality of this collection of articles. Last, but not least, our gratitude goes to Gabriela Wagner, whose editorial assistance and handling of all the communication with the authors and the reviewers finally made this volume possible.

October 2015

Hendrik Decker
Lenka Lhotska
Sebastian Link

Organization

Editorial Board

Contents

Reflective Constraint Writing
A Symbolic Viewpoint of Modeling Languages

Dirk Draheim$^{(\boxtimes)}$

University of Innsbruck, Innsbruck, Austria
draheim@acm.org

Abstract. In this article we show how to extend object constraint languages by reflection. We choose OCL (Object Constraint Language) and extend it by operators for reification and reflection. We show how to give precise semantics to the extended language OCL_R by elaborating the necessary type derivation rules and value specifications. A driving force for the introduction of reflection capabilities into a constraint language is the investigation of semantics and pragmatics of modeling constructs. We exploit the resulting reflective constraint language in modeling domains including sets of sets of domain objects. We give precise semantics to UML power types. We carve out the notion of sustainable constraint writing which is about making models robust against unwanted updates. Reflective constraints are an enabler for sustainable constraint writing. We discuss the potential of sustainable constraint writing for emerging tools and technologies. For this purpose, we need to introduce a symbolic viewpoint of information system modeling.

Keywords: Meta modeling · Multi-level modeling · Object constraint languages · Generative programming · Database migration · Schema evolution · clabjects · Modeling tools · UML · OCL · Z · GENOUPE

1 Introduction

An object-constraint language is a logical language that is embedded into a modeling framework and offers language constructs specific to object-oriented modeling. In this article we show how to add reflection to object-oriented constraint languages. Reflection is about access to the meta level, both introspective as well as manipulative. We need a reflective constraint language to analyze issues and express results in the semantics and pragmatics of information system modeling. Reflective constraints are an enabler for sustainable constraint writing, which is about making models robust against unwanted updates [29]. More specifically, we can exploit a reflective constraint language for:

- *Semantics of modeling languages.* Given meta-level access you can give precise semantics to existing modeling language constructs. We do this for UML power types in this article. Furthermore, you can use a reflective constraint language to extend an existing modeling language with new well-defined modeling constructs.

© Springer-Verlag Berlin Heidelberg 2016
A. Hameurlain et al. (Eds.): TLDKS XXIV, LNCS 9510, pp. 1–60, 2016.
DOI: 10.1007/978-3-662-49214-7_1

- *More adequate system analysis.* With today's technologies, i.e., databases, third-generation programming languages and modeling tools, we encounter a model-object divide. This model-object divide is not accidental; it is just a property of current mainstream information system technology, which is established and mature. Nevertheless, the model-object divide sometimes hinders us from stating fully adequate models of domain knowledge. This is so, because a model and its objects together intend domain objects and together encapsulate domain knowledge. For example, you might have some classes $A_1, \ldots, A_n$ and have found some constraints for these classes. Now, you might encounter that these constraints are instances of a general constraint pattern that must hold for an arbitrary number of classes. Without appropriate reflective features, you can only state such constraint patterns in the informal comments. A reflective constraint language is the solution for this. In general, we need full reflective support – limited forms of reflection, like generic types, are not sufficient.
- *Quality assurance for system design.*
  - Ensuring that class names follow a given style guide.
  - Ensuring that each attribute has correctly typed setter- and getter-methods.
  - Ensuring a complex design pattern.
  - etc.

All of the above items are practically motivated [20], i.e., reflective constraint writing is to constraint writing what generative programming is to programming. However, reflective constraint writing is also of importance beyond immediate practical exploitation. It can help in mitigating gaps between different information system paradigms. It can help in mitigating gaps between different viewpoints in information system modeling. We proceed as follows. We choose the OMG standard meta-level architecture as the backbone for our efforts. We extend the OCL (Object Constraint Language) with reification and reflection, resulting in the so-called OCL_R in Sect. 2. We show how to give a declarative semantics for OCL_R in Sect. 3. We review some OCL_R examples in Sect. 4 and also provide a comparison with generative programming, based on the concrete programming language GENOUPE.

In Sect. 5, we exploit OCL_R to specify constraints needed in modeling of sets of sets of domain objects. We streamline the discussion by showing how usual class diagrams, i.e., without multilevel modeling constructs, are sufficient to adequately model sets of sets of domain objects if appropriate constraints are provided and if and only if these are made robust against M1-level updates. We further streamline the discussion by considering sustainable constraint writing in the specification language Z in Sect. 6. Then, we generalize the found constraints further to give a precise semantics for UML power types 7. From these discussions, we extract more general notions like sustainable constraint writing and a symbolic viewpoint on modeling languages. In Sect. 8 we discuss model evolution, notions of constraints and viewpoints onto modeling languages. We discuss related work throughout the paper and summarize related work in Sect. 9. We end the article with a conclusion in Sect. 10.

In the Appendices A, B and C we provide overviews of the abstract syntax of the UML core language, the OCL v2.0 types and the OCL expression language.

2 OCL_R – A Reflective Extension of OCL

The OCL (Object Constraint Language) is syntactically and semantically embedded into the UML meta-level architecture. The aim of this section is to extend OCL with full reflection. Note, that we use the 2006 version of OCL, i.e. OCL v2.0 [70] as the basis for our language extension. We do neither use the current version OCL v2.4 [73] nor the version OCL v2.3.1 [72], which has been released as ISO standard ISO/IEC:19507 [51]. The reason for this is the particularly mature and precise definition of the OCL type system in the former version OCL v2.0 – see Appendix B for a discussion of this issue. If you need to delve into some of the concepts used in the upcoming sections, e.g., the OCL type *OclType* and its generating class *TypeType*, it is important that the standard v2.0 [70] is the authoritative reference for this article and not the newer standards. The choice of standard is for technical reasons only and not due to essential differences. For example, the abstract syntax of the OCL versions v2.0 and versions v2.3.1 and v2.4 are exactly the same. All crucial arguments and statements on OCL in this article, e.g., with respect to expressive power, are independent of the chosen standard.

- Properties of all objects, i.e., objects $o : OclAny$:
  - o.oclIsTypeOf(t:OclType):Boolean – *true iff* $o : t \wedge \nexists t'.t < t'$)
  - o.oclIsKindOf(t:OclType):Boolean – *true iff* $o : t$
  - o.oclInState(s:OclState):Boolean – *Test for state machine state.*
  - o.oclIsNew():Boolean – *Postcondition test for object creation.*
  - o.oclassType(t:OclType):instance of Classifier – *Type casting operation.*
- Properties of meta objects $t : OclType$ representing user-defined types:
  - t.name:String – *The name of the type t.*
  - t.attributes:Set(String) – *The set of names of the attributes of t.*
  - t.associationEnds:Set(String) – *Names of association ends navigable from t.*
  - t.operations : Set(String) – *The names of the operations of t.*
  - t.supertypes : Set(OclType) – *The set of all direct supertypes of t*
  - t.allSupertypes : Set(OclType) – *The set of all supertypes of t*
  - t.allInstances : Set(type) – *The set of all instances of type t.*

Fig. 1. OCL inbuilt meta-level access.

2.1   Meta-object Access in OCL

Standard OCL offers only limited access to the meta-level. The complete list of these OCL meta-level access operations is given in Fig. 1. The OCL meta-level access is restricted to introspection, i.e., no constraint generation is supported. Even the introspective features are limited. First, way not all meta-relationships that are established by the UML meta model have a counterpart in the OCL language. Second, and this is actually the crucial point, the entry to the introspection is only in terms of the current context of an OCL expression and therefore in terms of only a fixed number of constantly defined user-types. This means, OCL's meta access capabilities yield no functional abstraction over user-defined types and therefore do not add to the expressive power of OCL. The meta access of OCL shows in properties for meta objects representing user-defined types, i.e., objects of type *OclType* and properties that expect a parameter of type *OclType*.

With respect to properties for meta objects, the property *allInstances* is the only one that is specified in the OCL standards since version v2.0. All the other stem from the first version v1.1 [69]. The following list of example constraint expressions cannot be expressed with the OCL inbuilt meta-object access capabilities – please compare the list also to Fig. 1:

1. Names of subclasses of a given type t.
2. The subclasses of a given type t.
3. Attribute names of classes navigable via associations from a given type t.
4. All classes of the user model.
5. The number of classes in the user model.
6. All classes of the user model that have no subclasses.
7. The sum of all Integer attributes of all objects of all classes.
8. Test, whether all attributes of all objects of all classes are initialized.
9. Test, whether all attributes of all classes have setter- and a getter-methods.

All of the above constraint expressions (1) through (9) can be expressed by the OCL-extension OCL$_R$. The several constraint expressions express different levels of sophistication. The first two constraints (1) and (2) could be made possible by augmenting the list of inbuilt OCL expressions in Fig. 1 by appropriate properties. However, in order to enable all the other constraints a more conceptual refactoring of OCL is necessary, because they long not only for introspective access but also for reflection.

The reflective programming language community distinguishes between *reification* and *reflection* – see also Table 1. Reification turns information about a program, i.e., meta-data, into data and makes it accessible to the programming level. Then, reflection can be understood as the exploitation of reified data. We then also talk about reflection in the wider sense. Reified data can be exploited in two ways. First, it can be exploited for introspective access. Second, it can be used to manipulate program structures. The reified data can be turned into program code itself, we then say that reified data is materialized or re-materialized. We then also talk about reflection in the narrow sense. The usual word for the materialization step of turning reified data into code is generation. We feel that generation somehow stresses more the operational facet of this mechanism. We use both generation and materialization as equal terminology. This terminology works also with respect to meta-level access in modeling languages and also with respect to constraint writing. Here, reification is about making meta-data accessible to the modeler. Again, reification allows for access to the meta-level and can be exploited for introspection and reflection in the narrow sense, i.e., materialization of reified data into modeling elements.

Table 1. Attempt to summarize some important reflective programming terminology and its application to reflective constraint writing.

Reflective programming. (Reflective Constraint Writing). Reflective Languages.	• Reification • Reflection (in the wide sense): Exploitation of Reified Data	• Introspection. • Reflection (in the narrow sense): materialization (generation) of reified data into code (constraints).

The OCL meta-level access offers only a limited form of reification. The OCL standards [72] explicitly state that OCL does not support the reflection capabilities of the MOF (Meta Object Facility) [76]. Note, that it is not sufficient to add syntactical constructs to a language like OCL to support reflective features. The real work lays in the elaboration of the semantics of such reflective capabilities as, e.g., provided by OCL_R. Shallow statements of the intended meaning of syntactical constructs would not be sufficient as semantic elaboration.

2.2   On the Chosen Declarative Approach for OCL_R Specification

Without loss of generality, we will define OCL_R as an M1-level language, i.e., we define the reification operators Φ and Ψ as well as the concrete syntax $\langle \_ \rangle \downarrow$ and $\langle \_ \rangle \uparrow$ used for them against the background of writing M1-level constraints. Similarly, we specify the well-formedness rules and the semantics of OCL_R from the perspective of writing M1-level constraints. Writing M1-level constraints is the major use case of OCL_R. Writing M1-level constraints is about adding constraint expressions at level M1. For OCL this means that writing M1-level constraints is about writing constraints for M0-level objects. With OCL_R it is possible to write meta object constraints, in particular, constraints on user-defined types at level M1 and therefore extend the semantics of meta models. We will see the specification of the UML power types semantics in Sect. 7 as an example for this. Therefore, there is no need to explicitly generalize the current definitions from a M2-level perspective.

2.3   On the Preciseness of the Chosen Specification Approach

We show how to provide a precise semantics description of OCL_R in this article. In the given OCL_R definitions we rely on the existing UML and OCL semantics defined in [70, 74] as the foundation for our semantic extensions. Note that our definitions are *free over* the semantic definitions yielded by the OMG specification. This means that even if semantic definitions in the OMG stack might be ambiguous or underspecified for some points, our semantics does not suffer. Furthermore, our specification varies in the way semantic decisions are made for the UML stack. There is no need for us to re-formalize or to fix UML semantics. We can simply assume the semantics as completely specified. The OMG stack forms a sweet spot between preciseness and convenience; at least, the core of it has a widely known and accepted semantics. See also [85] for a discussion of UML 2.0 semantics.

2.4 On Abstract Syntax Oriented Reflection

A reflection mechanism can have a design that is oriented throughout towards abstract syntax or, what we call, an ASCII-based design. In an ASCII-based design meta data is reified as text, i.e., 'String' data. Then reflection operators craft model elements from 'String' data input. In a thoroughly abstract syntax oriented design the data type of reified data is kept abstract and reflection is also realized by operations on this abstract data type. An abstract-syntax oriented design offers an important advantage. It makes it much easier to give precise semantics to the reflection mechanism, in particular, with respect to level-crossing type safety. With an ASCII-oriented design it is easier to provide ad-hoc implementations for a reflection mechanism, in particular, it the implementation has to be provided for an existing platform. The design of OCL_R is thoroughly oriented towards abstract syntax.

Fig. 2. Class diagram.

2.5 Notational Issues of OCL Contexts and OCL Meta Objects

As a minor issue we sometimes want to get rid of context notation in constraint writing in the sequel. The concepts of contexts and meta objects, i.e., objects that represent types, are completely exchangeable. First, consider the following OCL invariants, which are written against the tiny class diagram in Fig. 2:

$$\textbf{context } \textit{Person} \textbf{ inv: } \textit{age} \geqslant 40 \tag{1}$$

$$\textbf{context } \textit{Foo} \textbf{ inv: } \textit{Person.allInstances} \rightarrow \textit{forAll}(\textit{age} \geqslant 40) \tag{2}$$

It is easy to see that the constraints (1) and (2) have the same semantics. Now, we can see that the role of *Person* in (1) and (2) are exactly the same. On the on hand, in (1), you can consider *Person* a meta object. One the other hand, in (2), you can consider *Person.allInstances* $\rightarrow$ *forAll*(_) to provide context for the evaluation of *age* $\geqslant$ 40. Consider the following constraint:

$$\textit{Person.allInstances} \rightarrow \textit{forAll}(\textit{age} \geqslant 40) \tag{3}$$

The information in constraint (3) is complete. The type *Foo* in (2) is not needed in the subsequent expression, it is merely a wildcard, so it can be dropped to yield (3) without loss. Henceforth, we will often write constraints without explicitly given context, in the style of (3). Though the constraint in (3) is clumsier than its version in (1) it is easier to handle in formal argumentations like type derivation or value specifications. Note, that the concrete syntax in (1) and (2) is official OCL syntax; although it is rather used in standard documents as opposed to the more embellished concrete syntax usually found in textbooks.

2.6 Terminology for the OMG Meta-level Architecture

We need to introduce some notation and terminology for issues in meta modeling architectures to be used in the sequel. The introduction of these notations must not be misunderstood as an attempt to specify, or let's say better, to re-specify the UML meta level architecture and its languages. We take the standard OMG four-level meta model hierarchy, see [74, Sect. 7.12], as background architecture, see also Fig. 3. Syntax and semantics of the UML meta model, the UML meta model and OCL are taken as granted as defined in [70, 72, 74–76].

However, the concepts introduced in this section go beyond mere notational issues. We also define important terminology, hand in hand, with notation for it. This way we define the value identity for objects in the meta-level architecture. This value identity is defined across the levels of the meta-level architecture, i.e., it is introduced to make objects at different levels of the architecture comparable. Based on the value identity we will define the meta model reification operator Φ and the model reification operator Ψ.

UML Meta Model Notation. We denote the UML meta model by $\mathfrak{M}_2$. Similarly, we denote the UML meta model by $\mathfrak{M}_3$.

Object Notation. We denote the set of all primitive values by P. The set P is flat, i.e., it is the union of all interpretations of UML's primitive types. We introduce a set of object identifiers and denote it by OID. We denote the set of all attribute values by $V = \mathbb{P}(P \cup OID)$. The power set in the definition of V is necessary, because the UML attributes are, in general, many-valued. We denote the set of attribute names or labels by L. We denote the set of finite subsets of a set M by $\mathbb{F}(M)$. Conceptually, in our notation, an object consists of an object identifier, a finite set of labels from L and an attribute value for each of these selected labels. We define the set of all objects O as Cartesian product of object identifiers OID and finitely L'-indexed sets of attribute values, for all possible subsets of labels, i.e.:

$$O = OID \times \bigcup_{L' \in \mathbb{F}(L)} (V_l)_{l \in L'} \tag{4}$$

The way we defined O, objects are denoted as records [1,17], or to be precise, object values are denoted as records, and objects are formed as an object reference to record. We use the usual notation for records, i.e., $\langle oid \mapsto \langle l \mapsto x_l \rangle_{l \in L} \rangle$.

O contains many objects that are impossible, i.e., objects that can never be instances in the UML meta level hierarchy. This is so because our objects are completely untyped assemblies. They are based on the value set V which is completely flat. This does not harm, because the definitions in Sect. 2.6 are not about semantics, but about notation. We will ensure the well-formedness of object and models of OCL_R later by the definition of the typing relationship $\_ :\_$ and the instantiation relationship $\_::\_$ for the corresponding extensions to OCL.

Note that O is the full extension of the meta level hierarchy, i.e., the collection of all potential objects that can be materialized in system states. The set O is a forgetful viewport. It only models aspects that are needed in the upcoming semantic definitions. For example, it forgets ordered association ends. In O we combine information on primitive-typed attributes with object references into a record. Another possibility would have been to denote meta level elements as records of merely primitive-typed values plus explicit object links as second kind of instances. Note, by the way, that in the UML semantics both styles of element presentations redundantly co-exist – see Fig. 4, diagram (v). The third option is to represent elements as pure nets of object identifiers with primitive values as leaves. By the way, we have discussed the latter option in form-oriented analysis as so-called parsimonious data model [26, 37]. Once more note, that the purpose of Sect. 2.6 is not to formalize UML semantics. It is merely about establishing notation for the existing meta level framework to be exploited in upcoming sections.

We have designed the value space as $V = \mathbb{P}(P \cup OID)$. As we have said, in the UML an attribute is, in general many-valued. Only, in the special case that the cardinality of an attribute is 1..1, an attribute is single-valued. The standard evaluation of an attribute in UML yields a bag, not a set. We have not designed our values in V as bags but as plain sets. This does not pose a problem, because bags can be formed by exploitation of object references. In the UML, properties can also evaluate to sequences. We assume that this sequencing can be modeled by an indexing mechanism on labels. We are interested in keeping our notation as reductionist as possible.

Meta-object Levels. We use $O_i \subset O$ to denote the set of all objects at meta-level M_i, the M_i-level objects for short.

Instances. Given an M_i-level object o and an M_{i+1}-level object C, we use $o :: C$ to denote the fact that o is an instance of C as defined by the UML specification. Given an object $o \in O$ and a set of objects $M \subset O$, we use $o :: M$ to denote the fact that there exists a $C \in M$ so that $o :: C$. Given sets of objects $M, N \subset O$, we use $M :: N$ to denote the fact that $o :: N$ for all $o \in M$. In case that $M :: N$ we also say that M is a instantiation of N.

Models. We call a subset $m \subset O$ of objects a model **iff** m is a partial function, i.e.:

$$m \in OID \rightarrow \bigcup_{L' \in \mathbb{F}(\mathfrak{L})} (V_l)_{l \in L'} \tag{5}$$

Given a model $m \subset O$ we say that m is a model at level i, or M_i-level model for short **iff** for all $o \in m$ we have that $o \in O_i$. Given models $M, N \subset O$ we say that M is a model of N **iff** $M :: N$.

Value Identity. Next, we define *value identity* of objects with respect to given models. Given models m, n, an object $o \in m$ with $o = \langle oid \mapsto \langle i \mapsto x_i \rangle_{i \in I} \rangle$ and an object $p \in n$ with $p = \langle pid \mapsto \langle j \mapsto y_j \rangle_{i \in I} \rangle$, we define o and p to be value-identical, denoted by $o \equiv p$ **iff** for all attribute labels $i \in I$ we have that:

$$
\begin{aligned}
&(i) \quad x_i \not\subseteq OID \Rightarrow (x_i = y_i) \\
&(ii) \quad x_i \subseteq OID \Rightarrow \left(\exists \beta : x_i \leftrightarrow y_i \; . \; \forall x' \in x_i \; . \; m(x') \equiv n(\beta(x')) \right)
\end{aligned}
\tag{6}
$$

The definition of $\equiv$ is a partial specification only. It is only defined for objects that share the same set of labels I. It is only complete for well-typed and at the same time identically typed pairs of objects. This does not harm, because in the sequel we only work with well-formed models. Value identity can be characterized as identity up to exploited object references. The abstraction from concrete object references is exactly what is achieved by the bijection β in (6). In terms of programming languages, e.g., in Java terminology, value identity results from deep copying or cloning an object net.

Meta Model Embedding. We define the *embedding* of the UML meta model into the UML meta model $\iota : \mathfrak{M}_3 \hookrightarrow \mathfrak{M}_2$ by $\iota = \{(x, y) \mid x \equiv y\}$ – see also Fig. 3.

Standard Notation for Functions. For the sake of completeness, we recap some standard notation for functions. Given a function $f : A \rightarrow B$, we denote the *lift* of f by $f^\dagger : \mathbb{P}(A) \rightarrow \mathbb{P}(B)$, which is defined as usual. Given a function $f : A \rightarrow B$ we denote the reversal, as usual, by $f^{-1} : B \rightarrow \mathbb{P}(A)$.

Meta Model Reification. Next, we introduce the meta model reification operator Φ. First, we define the set of all meta model reification operators $\boldsymbol{\Phi}$ as the set of embeddings $\phi : \mathfrak{M}_2 \hookrightarrow O_1$ for which it holds true that (i) $\phi(\mathfrak{M}_2)^\dagger :: \mathfrak{M}_2$ and (ii) for all $m \in \mathfrak{M}_2$ it holds true that $\phi(m) \equiv m$. Then, we define Φ as an arbitrary but fixed element of $\boldsymbol{\Phi}$, i.e., $\Phi \in \boldsymbol{\Phi}$. In the sequel, we refer to Φ as *the* meta model reification operator. Note, that $\phi(m) \equiv \phi'(m)$ for all $\phi, \phi' \in \boldsymbol{\Phi}$ and $m \in \mathfrak{M}_2$. Furthermore note, that $|\mathfrak{M}_2| = |\phi^\dagger(\mathfrak{M}_2)|$, because ϕ is an embedding. This means, that all $\phi, \phi' \in \boldsymbol{\Phi}$ can be characterized as identical up to exploitation of object references. This explains, why it makes sense to define ϕ as an arbitrary but fixed selected element of $\boldsymbol{\Phi}$. In any case, formally, the definition based on a selection is well-defined.

Model Reification. On the basis of the meta model reification operator Φ we introduce the model reification operator Ψ. We define the set of model reification operators $\boldsymbol{\Psi}$ as the set of embeddings $\psi : O_1 \hookrightarrow O_0$ for which it holds true that, given any M_1-level model $\boldsymbol{m} \subset O_1$, it holds that (i) $\psi(\boldsymbol{m})^\dagger :: \Phi(\mathfrak{M}_2)$ and (ii) for all $m \in \boldsymbol{m}$ it holds true that $\psi(m) \equiv m$. Again, we define Ψ as an arbitrary but fixed element of $\boldsymbol{\Psi}$, i.e., $\Psi \in \boldsymbol{\Psi}$. In the sequel, we refer to Ψ as *the* model reification operator. Informally, the effect of the meta model reification Φ is to copy the meta model to the model level, whereas the effect of R the model reification Ψ is to copy the *user user* model to the object level. Take a look at Figs. 3 and 4 for a visualization of how this actually works.

Further Notational Issues. We model bags as functions to the ordinals, i.e., given a set T, we model the bags $Bags(T)$ of T as $Bags(T) = T \to \mathbb{N}_0$. Given a set T, we model the sequences $Seq(T)$ of T as indexed sets $(s_i)_{i \in \{1,..,n\}}$ over T with respect to a starting fragment $1,..,n$ of the ordinals. We define the length of a sequence as $\#((s_i)_{i \in \{1,..,n\}}) = n$. We use also $\lambda\, i \in \{1,..,n\}.s(i)$ to denote a sequence in $Seq(T)$.

2.7 Reification for Constraint Languages

A straight-forward approach to extend OCL by introspective and reflective features was to rewrite its syntax and semantics by doubling terms for the different levels of the meta-level architecture. Instead, we choose an economically approach that allows us to let the semantics of OCL almost untouched. We will have to give well-formedness rules and semantic specifications only for the newly added, genuine OCL_R reflection expressions. We achieve this by preparing the M1-level with a reified version of the UML meta model and the M0-level with a reified version of the user model – see Fig. 3.

The operator Φ reifies the UML meta model at level M1. Basically, this reification amounts simply to copying the UML language specification as a class diagram to the user level. This is immediately possible because of the bootstrap approach of the UML specification, i.e., because the UML meta model is specified in a core language that is itself a part of the UML language. Intuitively, we can say that we use the operator Φ to copy the UML meta model and add it to the user-defined model at level M1. Actually, the definition of Φ as provided in

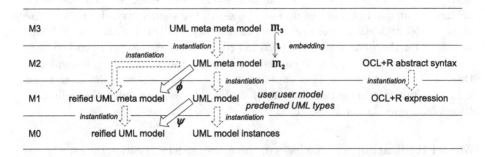

Fig. 3. Extending OCL with reification and reflection.

Sect. 2.6 is completely declarative. We have defined the set of object references O as an abstract data type. We keep O completely opaque, i.e., we do not define operations for the creation of object handles or the construction of objects. The value identity $\equiv$ that we have defined in (6) is a structural equality of object nets up to object references. Now, also the definition of Φ is free from concrete object construction mechanisms. We can assume the existence of Φ and therefore all semantics definition in this section are founded, in particular, the typing rules. If you find it helpful, you can think of the act of selecting and fixing an arbitrary ϕ from Φ as the act of copying model $\mathfrak{M}_2$ to level M1.

Each UML meta model expression is therefore immediately a correct M1-level model expression. This fact is also indicated by the embedding $\iota : \mathfrak{M}_3 \hookrightarrow \mathfrak{M}_2$ of the UML meta meta model into the UML meta model. Figure 3 shows the overall scenario of reification and reflection with OCL, whereas Fig. 4 gives a concrete example, based on a small cutout of the UML superstructure specification and a tiny user model. After the addition of the reified meta model to level M1 it is actually really a part of the user model. This fact eases the introduction of new reflective features to the OCL. However, usually we want to distinguish the reified meta model from the model that is actually created by the M1-level user modeler for its genuine purpose, e.g., domain modeling, system analysis, system design, and so forth. Henceforth, we call this part of the user model the *user user model* in cases where disambiguation seems to be important – see Fig. 4. The reification of the meta model data has to be understood as a semantic device, i.e., a means to declare the semantics of the extended language OCL_R. Therefore, by definition, there is no conflict with other software artifacts. The target of this article is not to achieve a particularly smart constraint language – whatsoever the criteria might be with respect to this. We add reflection to a constraint language for conceptual purposes. Ease and preciseness of the semantics are the rationales of the proposal. We are interested in the possibility of introducing reflection to object constraint languages in general. The resulting reflective language is interesting in its own right, but is not the ultimate goal.

With the reification of the UML meta model at level $M1$ we are prepared for introspective access. Given a meta model type, i.e., an $\mathfrak{M}_2$ type T, we use the concrete syntax $\langle T \rangle \downarrow$ to denote its reification at level M1. The $\langle \_ \rangle \downarrow$ notation is needed to distinguish user-defined types from reified meta model types. For example, if you want to model the national school system you might want to have a class *Class* in your model, and this class must not come into conflict with the reified meta model type *Class*. With UML, this disambiguation of types is not only an issue of the concrete syntax but also an issue of the abstract syntax. According to the UML superstructure, the name of a named element allows to identify the element unambiguously – see [75, Sect.7.3.34]. This means that the UML offers not a completely abstract modeling backbone. Therefore, the concrete syntax $\langle T \rangle \downarrow$ stands for opening of a namespace. In practice, we can get rid of the extra notation. We can simply assume that the namespaces of user-defined and reified types are separated and use names T of reified types $\langle T \rangle \downarrow$ without harm. Nevertheless, in this article we stay with the notation $\langle \_ \rangle \downarrow$ for reasons of preciseness and clarity.

Now, we step further by reifying the *user user* model at level M0. Because of
Φ, the appropriate classes of the reified UML meta model are available for this
purpose at level M1. The operator Ψ re-instantiates each model element $e_1 :: e_2$
as the value identical model element $\Psi(e_1) :: \Phi(e_2)$. What we have achieved now
is full introspective access onto the user-defined types, even without extension of
the OCL syntax. Again, we use the notation $\langle T \rangle \downarrow$ to denote the reification of a
user-defined type T, i.e., a type of the *user user* model. Note, that we overload
the notation $\langle \_ \rangle \downarrow$ to denote both Φ- and Ψ-reifications.

See once more, how the copying mechanism of the reification operators Φ and
Ψ work in Fig. 4. The copying step from sub diagram (*iii*) to sub diagram (*v*)
seems to unfold the diagram (*iii*). However, it does not. Diagrams (*iii*) and (*v*)
are just alternative visualizations of the same, i.e., value identical, object net,
where diagram (*v*) is of course more detailed. The usual class diagram notation
of (*iii*) is convenient for us, in particular, if we want to conceive it in its role as
an O_1-level model for the instantiation of M_0 level models like the sub diagram
(*vi*). However, diagram (*iii*) is an object net, and in its role as an instance of the
UML meta model $\mathfrak{M}_2$ we would perhaps want to perceive it rather as detailed
as visualized in (*v*).

2.8   Reflection for Constraint Languages

This section provides some examples of OCL_R expressions and an informal
description of their semantics. In Sect. 3 we show how to give precise semantics
in terms of type derivation rules and value specifications. With the meta model
and model reification operators Φ and Ψ in Sect. 2.7 we have already achieved
full introspective access to the *user user* model. However, yet the crucial step
is missing, i.e., gaining fully reflective access onto model elements of the *user
user* model via the reified data. What is missing is a means of materialization or
re-materialization of modeling elements, i.e., of reflection in the narrow sense –
see Table 1 once more. Many interesting constraints are yet not possible to write.
To get the point, we will look at a series of example constraints. The examples
serve merely as demonstration of the OCL and OCL_R mechanics. They are not
meant to present examples of domain knowledge. In later sections we will see and
discuss many exploitations of the reflective features that we have added to the
constraint language. For the purpose of easy reference, we have added a crucial
chunk of the UML meta model in Appendix A, a specification of the OCL type
system in Appendix B and a crucial chunk of the OCL abstract syntax specification in Appendix C. For many of the upcoming examples it will be helpful to
switch between the text and these Appendices back and forth. We start with the
following, correct constraint that exploits the reified data[1]:

$$\langle Class \rangle \downarrow .allInstances \rightarrow forAll(ownedProperty \rightarrow asSet \rightarrow size \leqslant 20) \qquad (7)$$

---

[1] Note, that we feel free to drop brackets from OCL operation calls whenever the
paramter list is empty, e.g., we write $s \rightarrow asSet \rightarrow size$ instead of $s \rightarrow asSet() \rightarrow size()$.

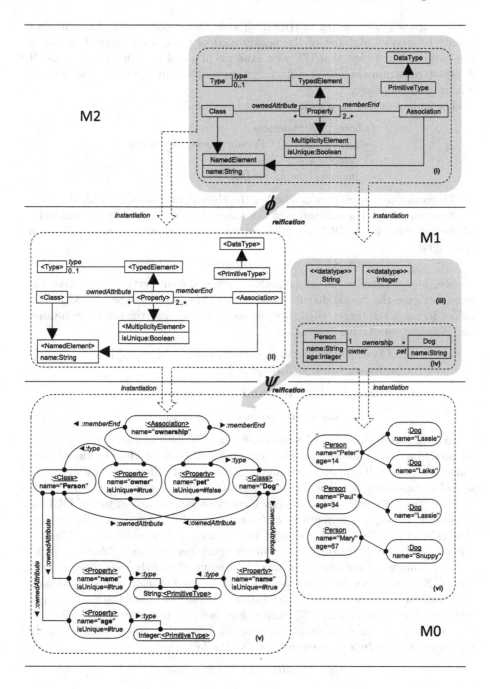

Fig. 4. The OCL$_R$ reification mechanism

Constraint (7) evaluates to true if all *user user* model classes have at most twenty properties. Against of the background of today's established modeling practice it is fair to say that (7) is an example of a real meta level constraint. It does not constrain the object during system evolvement time, but the modeler during modeling time.

Let us have a look at another constraint example:

$$\begin{aligned}
&\langle Class \rangle \downarrow .allInstances \\
&\rightarrow select(name \doteq "Person").ownedProperty \\
&\rightarrow select(name = "pet").type \\
&\rightarrow includes(name = "Dog")
\end{aligned} \tag{8}$$

The constraint in (8) checks whether the class *Person* is associated, via a role *pet* to a class *Dog*. The semantics of the constraint (8) is equal to the semantics of the following usual OCL constraint that work without the new reification capabilities:

$$\textbf{context } Person \textbf{ inv}: pet.oclIsTypeOf(Dog) \tag{9}$$

See, how constraint (9) immediately queries the property *pet*, whereas (8) must navigate the two additional links *ownedProperty* and *type* of UML meta model to reach the target *Dog*. Now, let us have a look at the following invalid, i.e., ill-typed, constraint expression:

$$\begin{aligned}
&\langle Class \rangle \downarrow .allInstances \rightarrow select(name = "Person"). \\
&allInstances \rightarrow forAll(age \geqslant 40)
\end{aligned} \tag{10}$$

Intuitively, constraint (10) has the following semantics as (1), i.e.:

$$\textbf{context } Person \textbf{ inv}: age \geqslant 40 \tag{11}$$

In OCL_R we will be able to write constraints like (10) in due course, after the introduction of an appropriate reflection notation. However, for the time being, constraint (10) is ill-typed. The problem is the second *allInstances*-property. To see why, consider the following type derivation. The expression $\langle Class \rangle \downarrow$ has type *OclType*. As part of the limited meta data access capabilities of OCL, it is possible to apply the method *allInstances* to this expression. The expression $\langle Class \rangle \downarrow .allInstances$ has type $Set(\langle Class \rangle \downarrow)$. The expression $\langle Class \rangle \downarrow$ .allInstances $\rightarrow select(name = "Person")$ again has type $Set(\langle Class \rangle \downarrow)$, actually, it evaluates to the one-set element consisting of exactly the reified *Person* object. Now, when we try to invoke the *allInstances* method to this expression, we provoke a type error, because *allInstances* can only be applied to terms of type *OclType*. The constraint in (10) simply does not adhere to the well-formedness rules of OCL. As an even simpler counter example, it is not possible to apply *allInstances* twice in a path expression like the following:

$$\langle Class \rangle \downarrow .allInstances.allInstances \tag{12}$$

Again, the expression in (12) is not well-typed. Again, intuitively, constraint expressions (12) has a semantics. It is intended to mean the set of all instances of all classes of the *user user* model. Again, in OCL_R we will be able to write constraint expressions like (12) in due course.

2.9   Full Reflection Capabilities

Now, we introduce a reflection construct $\langle\_\rangle\uparrow$ as the crucial extension to OCL. Based on this notation, we can give a correct version of constraint (10) as follows:

$$\langle\!| \langle\langle Class\rangle\downarrow .allInstances \rightarrow select(name = "Person")\rangle\uparrow |\!\rangle. \\ allInstances \rightarrow forAll(age \geqslant 40) \tag{13}$$

Constraints (13) checks whether each instance of the class *Person* has an attribute value over 40 for its attribute *age*, i.e., it is equal to constraint (11). Informally, the semantics of the reflection construct is the reversal of reification. In (13) the reflection receives the one-element set containing the reified *Person* object and yields the one-element set containing the *Person* meta-object it has been reified from. More precisely, the reflection construct in (13) turns an expression of type $Set(\langle Class\rangle\downarrow)$ into an expression of type $Set(OclType)$. See how this works in the following example type derivation. Then, as usual, $e : T$ means that a sub expression e has type T – see also Appendix B as a reference for OCL types:

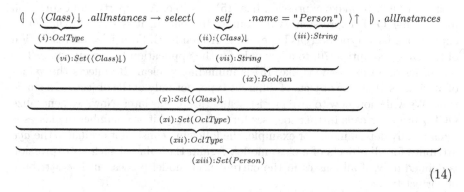

$$\tag{14}$$

The result of the reflection (xi) in (14) has type $Set(OclType)$. Unfortunately, with standard OCL this result cannot be immediately exploited in a property call *o.p*. OCL requires that a property can only be applied to an object of a single classifier – the OCL specification states [70]: *A PropertyCallExpression is a reference to an Attribute of a Classifier defined in a UML model. It evaluates to the value of the attribute.* We have two options to deal with this. We can extend OCL so that it can also deal with the application of a property to object of several classes and we will see in due course that this is easily possibly. As the second option, and this is what we see in the current example, is to introduce a new operator to OCL that turns a one-element set into the contained element. With $\langle\!| M |\!\rangle$ we denote exactly this operation. Note, that $\langle\!| M |\!\rangle$ is only partially defined, i.e., it is defined only for one-element sets. With respect to semantics, the solution based on $\langle\!| M |\!\rangle$ is conservative, i.e., it can be added to OCL without changing the existing semantics of the OCL.

With the reflection operator so far, we have added substantially to the expressive power to OCL. However, to earn the full potential reflective power, we need to develop a means to apply a property to a set of objects that are instances of completely arbitrary classes. In OCL we are already used to apply an attribute to a set of objects yielding an object set as result. In general, there is no reason why we should not apply an attribute to objects of more than one class. Have a look at the following example that introduces some ad-hoc concrete syntax for the special case of a fixed number of classes:

$$\textbf{context } \{Person, Dog\} \textbf{ inv: } self.age \geqslant 5 \tag{15}$$

Please note, that it is not our intention to introduce new concrete syntax. There is no need for us to do so. However, possibilities to address properties possessed by objects of more than one arbitrary type arise in OCL_R as a result of its design. Concrete syntax like the one in (15) has only the purpose to analyze the semantics of such scenarios for us. Of course, in (15) we assume that both the class $Person$ and the class Dog define Integer attributes age. Note, however, that the Integer attributes age in Dog and $Person$ are not required to be inherited from a common supertype of $Person$ and Dog – this is what we meant with $completely$ $arbitrary$ classes above. Fortunately, we do not need to change the syntax of OCL to make expression like (15) possible. A property call expression has another OCL expression as its source. In general, it is possible that OCL expressions have type Set(OclType) – see Appendix B. So it is a self-restriction of the OCL semantics [70] to allow only for the application to a single type. The semantics of an expressions like (15) is immediately clear. It collects the values of attributes of all the objects of different type, not only of the objects of a single type. We will show how to give precise semantics to this later. Now, we generalize OCL_R property calls further to cases, in which an arbitrary number of classes is dynamically determined. For example, the following constraint evaluates the age attribute for all objects of a class model. Again, the value of such an expression is defined only, if all classes in the current user-model possess an age attribute of correct type:

$$\langle\langle Class\rangle \downarrow .allInstances\rangle \uparrow .allInstances \rightarrow forAll(age \geqslant 40) \tag{16}$$

Next, we introduce the reification notation $\langle\_\rangle \downarrow$ also for $user$ $user$ model types. It hands over the reification operator Ψ to the modeler. The following example constraint, which is equivalent to (11), including the crucial type derivation shows how this works:

$$\langle\langle Class\rangle \downarrow .allInstances \\ \rightarrow select(\underbrace{self}_{(i):\langle Class\rangle \downarrow} = \underbrace{\langle Person\rangle \downarrow}_{(ii):\langle Class\rangle \downarrow}))\uparrow .allInstances \rightarrow forAll(age \geqslant 40) \tag{17}$$
$$\underbrace{}_{(iii):Boolean}$$

So far, each OCL expression of type $\langle Class\rangle \downarrow$ or collection type $Set(\langle Class\rangle \downarrow)$ can be made subject to reflection. In general, we will expand reflection to all kind of reified data. Let us have a look at the following example, which is again equivalent to (11):

$$
\begin{aligned}
Person.allInstances &\to forAll(\\
self.\langle \\
&(\\
&\quad \langle Class \rangle \downarrow .allInstances \\
&\quad \to select(\langle Person \rangle \downarrow).ownedAttributes \\
&\quad \to select(name = "age") \\
&) \\
\rangle \uparrow \\
&\geqslant 40 \\
)
\end{aligned}
\tag{18}
$$

In (18), the constraint expression inside the reflection construct yields a reified property. After application of the reflection, this property can then be called. See the following type derivation for the crucial part of (18):

$$
\boldsymbol{P} \equiv_{DEF} \underbrace{\langle Class \rangle \downarrow .allInstances \to select(\langle Person \rangle \downarrow)}_{(ii):Set(\langle Class \rangle \downarrow)}
$$

$$
\underbrace{\underbrace{self}_{(i):Person} .\langle (\underbrace{\overbrace{\boldsymbol{P}.ownedAttributes}^{(vi)} \to select(name = "age")}_{(iii):Set(\langle Property \rangle \downarrow)}) \rangle \uparrow}_{(iv):Set(\langle Property \rangle \downarrow)}
$$
$$
(vii):Integer
$$
\tag{19}

Technically, the property name p in a property call expression $o.p$ itself is not a proper OCL expression, in the sense that it does not have a type. This does not harm in the type derivation of (19). It is exactly the reflection construct that opens a context for typed expressions. See how the type of (v) in (19) is immediately consumed by the type derivation with rule (36) from Sect. 3, i.e., how the typing of (vi) is not needed in the type derivation.

As a next step, we can also generalize the semantics of property call expressions further, so that the application of a set of properties to a class or a set of classes becomes possible. See the following example showing again some ad-hoc syntax, with obvious semantics:

$$
\textbf{context } \{Person, Dog\} \textbf{ inv}: self.\{age, weight\} \geqslant 0
\tag{20}
$$

In OCL_R we can exploit such an extension to the OCL semantics in an expression like the following:

$$
\begin{aligned}
\langle \langle Class \rangle \downarrow .allInstances \rangle &\uparrow .allInstances.\langle \\
\langle Class \rangle &\downarrow .allInstances.ownedAttributes \\
&\to select(type = Integer) \\
\rangle &\uparrow .sum
\end{aligned}
\tag{21}
$$

The constraint (21) is well-formed with respect to all *user user* models. This is ensured by the clause *select*(*type* = *Integer*) which ensures that only type-correct property calls occur. The expression in (21) is a solution to expression (7) in the example list in Sect. 2.1, i.e., the sum of all Integer attributes of all objects of all classes. As we have mentioned before, we are free to omit all the special syntax for reification, reflection and also element picking, i.e., $\langle \_ \rangle \uparrow$, $\langle \_ \rangle \downarrow$ and $(\!| \_ |\!)$ in the application of the OCL_R, unless we do not need it to for the disambiguation of expressions. See how this simplifies expressions, e.g., for the expression (21):

$$
\begin{aligned}
&Class.allInstances.allInstances.(\\
&\quad Class.allInstances.ownedAttributes \\
&\quad \rightarrow select(type = Integer) \\
&).sum
\end{aligned}
\tag{22}
$$

Nevertheless, we stay with the explicit notation throughout the rest of the article, as we have said before, for the reason of preciseness.

3   On The Precise Semantics of OCL_R Reflection

The purpose of this section is to show how to give precise semantics to an object-oriented constraint language. In the definitions of this section we make extensive use of notation introduced in Sect. 2.6 and heavily rely on the concepts defined earlier, e.g., the reification and reflection operators Φ and Ψ. We define the necessary well-formedness rules as strict augmentations to the existing notion of UML and OCL type correctness.

3.1   Typing Notation and Semantic Bracketing

Given a UML, OCL or OCL_R expression e and type T, the *typing* $e : T$ expresses that e is well-typed and has type T. We use further usual notation from the type system community [17,58,79] to express well-typing. The statement $\vdash e : T$ holds if the typing $e : T$ has been derived, i.e., has been proven. Typing rules are expressed in the following manner:

$$
\frac{\vdash e_1 : T_1 \dots \vdash e_n : T_n \quad C_1 \dots C_m}{\vdash e' : T'}
\tag{23}
$$

Given that we have already derived typings $e_i : T_i$ and further conditions C_i hold true, a typing rule of kind (23) allows to derive typing $\vdash e' : T'$. There are no other typings than those that can be derived by typing rules. Typing rules are instances of well-formedness rules.

Furthermore, we use so-called semantic bracketing to define the value of expressions. Given an expression M we use $[\![M]\!]$ to denote its value. With semantic bracketing we mean the natural declarative technique to define the semantics as a recursive function along the structure of abstract syntax trees, i.e., the semantics of an expression $[\![e\ e_1 \dots e_n]\!]$ as a value $\boldsymbol{E}([\![e_1]\!], \dots, [\![e_n]\!])$ with $\boldsymbol{E}$ being a sufficiently precise semantic description. It is important to understand, that all definitions in this section, including typing rules and semantic equations are always in terms of abstract syntax trees, even if we use concrete syntax to denote them. Here, we again rely into the semantics of OCL. We assume a sufficiently precise semantics for OCL expressions that is available in our semantics specification, i.e., we assume that $[\![e]\!]$ is defined if e is a pure OCL expression.

3.2 Typing Rules and Values

Figure 5 shows the OCL_R meta model. Elements of the OCL are shown in grey, whereas the new language constructs are shown in black – compare to the OCL specification in Appendix C. The meta model elements of all of the three new language constructs implement the *OCLExpression* interface, i.e., they are proper OCL expressions that also receive types. A reflection expression refers to another OCL expression as its reflected expression. A reification expression refers to a type expression as its reified type. The types of OCL_R are the same as the types for OCL v.2.0 – see Appendix B Fig. 12.

We consider the semantics for three kinds of expressions that cover the full range of OCL_R semantics, i.e., type expressions, property call expressions and enumeration literal expressions. As explained in Sect. 2.8 the reification of the crucial $\mathfrak{M}_2$-model element *Class*, i.e., $\langle Class \rangle \downarrow$ has the type *OclType*. We repeat this as the following rule:

$$\frac{}{\vdash \langle Class \rangle \downarrow : OclType} \tag{24}$$

Furthermore we now, by the definition of Φ and Ψ that for all types t of the *user user* model, i.e., the data types of an M1-level model, we have that:

$$\frac{\vdash t : OclType \quad t :: Class}{\vdash \langle t \rangle \downarrow : \Phi(Class)} \tag{25}$$

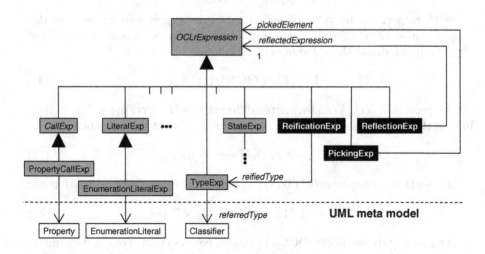

Fig. 5. Abstract syntax of OCL_R

Now, an expression of a reified data type, i.e., a reified data type expression, yields an *OclType*-expression after reflection. Reflection recovers the M1-level value of reified data:

$$\frac{\vdash e : \Phi(Class)}{\vdash \langle e \rangle \uparrow : OclType} \tag{26}$$

The value of a reified data type expression $\langle e \rangle \uparrow$ is defined as:

$$[\![\langle e \rangle \uparrow]\!] = [\![\, \Psi^{-1}([\![e]\!]) \,]\!] \tag{27}$$

In general, expressions rather have a type of the form $C(\Phi(Class))$ for a collection C than merely the type $\Phi(Class)$. Examples for expressions that have $\Phi(Class)$ as type are the iterator variables in OCL loop expressions. A further example is the result of correctly applying an element picking operation. The element picking operation can be applied to a collection of any type, including reified M2-level types. We define the typing and the value of the picking operator $(\!|\ \_\ |\!)$ for all kinds of collection C, i.e., *Set*, *Bag* and *Sequence* as follows:

$$\frac{\vdash e : C(T) \quad |[\![e]\!]| = 1}{\vdash (\!|\ e\ |\!) : T} \tag{28}$$

$$[\![\ (\!|\ m\ |\!)\ ]\!] = \begin{cases} [\![x]\!] & ,\ \exists x.m = \{x\} \\ \bot & ,\ else \end{cases} \tag{29}$$

Now, we generalize typing and values of reified data type expressions for the case of collection types. For all kinds of collections C, i.e., *Set*, *Bag* and *Sequence*, we have that:

$$\frac{\vdash e : C(\Phi(Class))}{\vdash \langle e \rangle \uparrow : C(OclType)} \tag{30}$$

With respect to the value of reified data type expressions we need to distinguish three cases now, i.e., sets, bags, and sequences. In case that $e : Set$ $(\Phi(Class))$, we define the value of $\langle e \rangle \uparrow$ as follows:

$$[\![\langle e \rangle \uparrow]\!] = [\![\ (\Psi^{-1})^{\dagger}([\![e]\!])\ ]\!] \tag{31}$$

We know that $[\![\langle e \rangle \uparrow]\!]$ can be written differently as $\{t : OclTpye \mid \Psi(t) \in [\![e]\!]\}$. In case that $e : Bag(\Phi(Class))$ we can define the value of $\langle e \rangle \uparrow$ as follows:

$$[\![\langle e \rangle \uparrow]\!] = \lambda t : OclTpye\ .\ [\![e]\!](\Psi(t)) \tag{32}$$

In case that $e : Sequence(\Phi(Class))$ we can define the value of $\langle e \rangle \uparrow$ as follows:

$$[\![\langle e \rangle \uparrow]\!] = \lambda i \in \{1, .., \#([\![\langle e \rangle \uparrow]\!])\}\ .\ \Psi^{-1}([\![e]\!](i)) \tag{33}$$

As a next step we specify OCL_R in case of enumeration types and enumeration literals. We define that, for all types t of the *user user* model, we have that:

$$\frac{\vdash t : OclType \quad t :: Enumeration}{\vdash \langle t \rangle \downarrow : \Phi(Enumeration)} \tag{34}$$

$$\frac{\vdash e : \Phi(EnumerationLiteral)}{\vdash \langle e \rangle \uparrow : EnumerationLiteral} \tag{35}$$

The value specification for enumeration literal expressions is identical to the one in case of type expression, i.e., Eq. (27).

3.3 Semantics of Property Call Expression

We turn to property call expressions now. First, we consider one of the simplest cases that (i) a reflected property is called on a single object, (ii) the reified property results into a single object and (iii) the property call results into a single object of user-defined type. In this case typing is defined for all user defined types T_2 as follows:

$$\frac{\vdash o : T_1 :: Class \quad p : \Phi(Property) \quad p.class = \Psi(T_1) \quad p.type = T_2}{\vdash o.\langle p \rangle \uparrow : \Psi^{-1}(T_2)} \tag{36}$$

In the scenario prescribed by (36) we define the value of a property call expression $o.\langle p \rangle \uparrow$ as follows:

$$[\![o.\langle p \rangle \uparrow]\!] = [\![o.\Psi^{-1}([\![p]\!])]\!] \tag{37}$$

For a full specification of OCL_R expressions we had to define a combinatorial number of different cases, depending on the result type of the property, the result type of reified operations and the question of whether the operation is applied to a single object or a collection of objects. Then, each of the involved types can be, combinatorial, a primitive type or a collection, and again, each collection, also the collection of objects, can be a set, a bag, or a sequence. We look at only one further case, which is a particular complex one, i.e., the case that all of the aforementioned components can be collections. For all kinds of collections C_1, C_1, C_3, i.e., Set, Bag or $Sequence$, we establish the following typing rule:

$$\frac{\vdash o : T_1 :: C_1(Class) \quad p : C_2(\Phi(Property)) \quad p.class = \Psi(T_1) \quad p.type = C_3(T_2)}{\vdash o.\langle p \rangle \uparrow : (C_1 \oplus C_2 \oplus C_3)(\Psi^{-1}(T_2))} \tag{38}$$

The definition of the typing rule (38) relies on a combinator $\oplus$ for collection constructors. This combinator is defined in Table 2 in Appendix B.2. The OCL approach is that nested collections are always and automatically flattened. For example, a set of sets is turned into a set, a bag of bags is turned into bag and so forth. The definition of the $\oplus$ combinator fulfills the standard definition of OCL collection flattening in [70,73]. First, we handle the case that all of the involved components yields sets, i.e., $C_1(Class) = Set(Class)$, $C_2(\Phi(Property)) = Set(\Phi(Property))$ and $C_3(T_2) = Set(T_2)$. In this case, we know that $o.\langle p \rangle \uparrow$ has the type $Set(T_2)$ and we define the value of $o.\langle p \rangle \uparrow$ as follows:

$$\llbracket o.\langle p \rangle \uparrow \rrbracket = \{v \in T_2 \mid v : \llbracket o'.p' \rrbracket, \; o' \in o, \; \Psi(p') \in \llbracket p \rrbracket\} \tag{39}$$

Next, we handle the case that all of the involved components yield bags, i.e., $C_1(Class) = Bag(Class)$, $C_2(\Phi(Property)) = Bag(\Phi(Property))$ and $C_3(T_2) = Bag(T_2)$. In this case, we know that $o.\langle p \rangle \uparrow$ has the type $Bag(T_2)$ and we define the value of $o.\langle p \rangle \uparrow$ as follows:

$$\llbracket o.\langle p \rangle \uparrow \rrbracket = \lambda v : T_2 . \sum_{o' \in o} \left(\sum_{p' \in \Psi^{-1}\llbracket p \rrbracket} \llbracket o'.p' \rrbracket(v) \right) \tag{40}$$

Note, that the sums in (40) are all well-defined, because all of the involved collections are finite. The scenario that the involved components yield bags, is the standard scenario in the OCL. We have started with the set scenario in (39) only for instructive purposes. We do not detail out further combinations $C_1 \oplus C_2 \oplus C_3$ of collection constructions.

4 Working with OCL$_R$

We will see OCL$_R$ at work in Sects. 5 and 7 when we exploit it for the semantic investigation of power types in general and power types in UML in particular. Before that, let us walk through the informal constraint examples that we have enumerated in Sect. 2.1 as constraints (1) through (9). Again, please have a look at the cutout of the UML meta model as provided by Fig. 11 in Appendix A throughout the examples. Example (1), i.e., the names of subclasses of a given type t can be expressed in OCL$_R$ as follows:

$$\begin{aligned} &\langle Generalization \rangle \downarrow .allInstances \\ &\rightarrow select(general = \langle t \rangle \downarrow).specific.name \end{aligned} \tag{41}$$

The subclasses of a given type t, i.e., example (2), follows immediately from (41) by dropping the last property call, i.e., the *name* navigation. Example (3), i.e., the attribute names of classes navigable via associations from a given type t can be expressed as follows:

$$\begin{aligned} &\langle Class \rangle \downarrow .allInstances \rightarrow select(c \mid \\ &\quad \langle Association \rangle \downarrow .allInstances \rightarrow exists(\\ &\quad\quad memberEnd \rightarrow contains(\langle t \rangle \downarrow) \\ &\quad\quad and \\ &\quad\quad memberEnd \rightarrow contains(c) \\ &\quad) \\ &).name \end{aligned} \tag{42}$$

Example (4), i.e., all classes of the user model, turns out to be a most simple example that has been exploited already in many instances before. It is given by $\langle Class \rangle \downarrow .allInstances$. Consequently, example (6), i.e., the number of classes in the user model is given by $\langle Class \rangle \downarrow .allInstances \rightarrow sum$. Example (6), i.e., all classes of the user model that have no subclasses, is provided by:

$$
\begin{aligned}
&\langle Class \rangle \downarrow .allInstances \rightarrow select(c \mid \\
&\quad not \\
&\quad \langle Generalization \rangle \downarrow .allInstances \rightarrow exists(\\
&\qquad general \rightarrow contains(c) \\
&\quad) \\
&)
\end{aligned} \tag{43}
$$

The example (7) has already been solved as an example by (21) before. A test, whether all attributes of all objects of all classes are initialized, i.e., example (8) can be realized as follows:

$$
\begin{aligned}
&\langle Class \rangle \downarrow .allInstances \rightarrow forAll(c \mid \\
&\quad \langle c \rangle \uparrow .allInstances \rightarrow forAll(o \mid \\
&\qquad c.ownedAttribute \rightarrow forAll(a \mid \\
&\qquad\quad o.\langle a \rangle \uparrow \neq null \\
&\qquad) \\
&\quad)
\end{aligned} \tag{44}
$$

Fortunately, in OCL a *null* value of type *OclVoid* is available as defined in [70,73]. This *null* value is exploited in (44) to test whether an attribute is initialized, where we assume that initialized attributes have a value different from *null*.

4.1   Getter and Setter Method Example

Next, we realize example (9), i.e., a test whether all attributes of all classes have setter- and a getter-methods. The following constraint (45) checks whether for each class and attribute X of type t there exist a setter-method and a getter-method of appropriate parameter signature, i.e., a method $setX(x : t)$ for some arbitrarily named input parameter and a method $getX() : t$:

$$
\begin{aligned}
&\langle Class \rangle \downarrow .allInstances \rightarrow forAll(c \mid \\
&\quad c.ownedAttribute \rightarrow forAll(a \mid \\
&\qquad c.ownedOperation \rightarrow exists(m \mid \\
&\qquad\quad m.name = "set" + "a.name" \; and \\
&\qquad\quad m.ownedParameter.size = 1 \; and \\
&\qquad\quad m.ownedParameter.direction = \#in \; and \\
&\qquad\quad m.ownedParameter.type = a.type \\
&\qquad) \; and \\
&\qquad c.ownedOperation \rightarrow exists(m \mid \\
&\qquad\quad m.name = "get" + "a.name" \; and \\
&\qquad\quad m.ownedParameter.size = 1 \; and \\
&\qquad\quad m.ownedParameter.direction = \#return \; and \\
&\qquad\quad m.ownedParameter.type = a.type \\
&\qquad) \\
&\quad) \\
&)
\end{aligned} \tag{45}
$$

Note, that the UML specification fixes that a method can have one and at most one return parameter. It would also be possible to completely specify the correct behavior of the methods based on the signature specification in (45) but we not detail this out here.

4.2   A Comparison with Genoupe Generative Programming

Reflective object-oriented constraint-writing is the natural counterpart to generative programming. Generative programming is another word for reflective programming. Generative programming gets its name from what we have called reflection in the narrow sense in Table 1, i.e., the step of turning reified data into code. Code generation is a particular operational viewpoint on reflection. It hints to a possible implementation strategy based on a pre-compilation phase for reflective features on top of an already existing programming language.

Now, as an instructive example, let us program the counterpart of the getter- and setter-example (45) in Sect. 4 in a reflective programming language. We choose our own reflective programming language GENOUPE [33,34,57] for this purpose – see Listing 4.1.

Listing 4.1. Generation of Getter and Setter Methods with GENOUPE

```
public class GetterSetter (Type T) : @T@{

    @foreach(F in T.GetFields()) {

        public void @"set" + F.Name@ ( x : @F.FieldType@) {
            @F.Name@ = x;
        }

        public  @F.FieldType@ @"get" + F.Name@ {
            return @F.Name@;
        }

    }
}
```

GENOUPE is an extension of the programming language C# with generative programming features. An important contribution of GENOUPE is the definition and implementation of an extended notion of *generator type-safety*, which is, however, not important for the consideration of the current example. For us, it is enough to understand how the generative features in Listing 4.1 work.

GENOUPE is extended by new, concrete syntax for meta programming. The special sign @ is used to introduce or embed some of the new meta-programming syntax. A pre-processing phase takes GENOUPE and generates plain C# code. In Listing 4.1 we implement a class *GetterSetter* parametric on a type parameter

(*TypeT*). Then, with :@*T*@ we achieve that the generated *GetterSetter* class extends the actual parameter class and therefore inherits all of the fields of the actual parameter class. In GENOUPE we have access to all features of the C# reflection API. In Listing 4.1 we exploit the method *T.GetFields()*, that yields all fields of a type *T*, as well as the properties *F.Name* and *F.FieldType* with respective meaning. Then, the GENOUPE meta programming expression @*foreach*($i\ldots$){$C(i)$} allows us to generate a piece of C# code for each instance of an iterator variable i. In Listing 4.1, we exploit @*foreach* to generate a getter and a setter method for each field of the actual parameter class.

5   Adequately Modeling of Sets of Sets

This section deals with the modeling of sets of sets of domain objects. Modeling sets of sets of objects is important, because it arises naturally in expert domains – see Fig. 6 for an example. Modeling of sets of sets is a classical topic [52] in the modelling community and has been discussed as modeling with power types [46,47,60,67,68]. It has also been discussed as multilevel modeling in the past [7].

Often, modeling a set of domain objects involves the specification of properties that are common to all objects of the investigated set. This means, that in domain modeling, we are, in general, also interested in the *intension* or *comprehension* of a set of objects rather than merely in its role as an *extension* of a concept. We could introduce new terminology as has happened in the object-oriented community in the past. For example, we could call a set of objects together with its intension a *class*. There is no single commonly accepted definition of the notion of *intension* in linguistics and ontology. So, let me be more concrete. More concrete, we could say that objects have properties and that we have a special interest into a certain notion, let's call it, e.g., *domain object class*, which is a set of objects together with a specification of which properties are shared by all objects of this set, i.e., are equal for all objects of that set. We could than give this notion a name and *class* has been a usual candidate for this in the past. The problem is that *class* is also used for concrete programming and modeling language constructs which usually have a rather operational semantics. Concrete class constructs in programming and modeling languages have been designed, of course, with a notion of *domain object class* in mind. In order to avoid conflicts with the class terminology of programming and modeling language constructs, we could choose another name for *domain object classes*, e.g. *Class, domain class*, or simply *domain object class*. We do not. We simply talk about set of objects, set of sets of objects and so forth and point out, that in domain modeling, we have special interest in properties that are common to all objects of a given set. The treatment of *intensions* of sets of objects have been intensively studied in the in the modeling community and many crucial results has been achieved for: notation, terminology, modeling constructs, patterns, tools, semantical considerations etc. – see also Sect. 9 for examples. Therefore, we have chosen this as an example worth looking at to analyze with a reflective constraint language like OCL$_R$.

As an example, Fig. 6 shows the mammal hierarchy. A set of sets appears systematically at each level of the hierarchy, grouping objects of classes that reside at a lower level, i.e., breed, sub species, species, genera and so on. The running example in this article will be dogs and breeds from this hierarchy. The different sets of sets in the mammal hierarchy in Fig. 6 all follow the *type-object pattern* [52] of Johnson and Woolf. The set of set is called a type in the type-object pattern. A type in the type object pattern, i.e., an instance of the type class of the pattern, is not a type in the sense of object-oriented subtyping hierarchy; in particular, it is not a type of a concrete modeling language. A type in the type object pattern represents a *kind* or a *group* of objects, i.e., a set of set of objects. Furthermore, a type carries the attributes common to all of the objects that it groups together. In that sense, again, a type of the type-object pattern is the *intension* of a set of objects.

The aim of this section is to discuss ways of adequately modeling sets of sets of domain objects. We show that it is possible to model sets of sets in terms of basic object-oriented modeling constructs plus appropriate constraints. Subtyping and subclassing mechanisms help in modeling sets of sets, but we will see that even most basic notions of object-oriented modeling, i.e., classes and enumeration types, are already sufficient for giving appropriate models of sets of sets, as long as the necessary, generic constraints are provided. There is no necessity to introduce new modeling language features, like a concept of set or

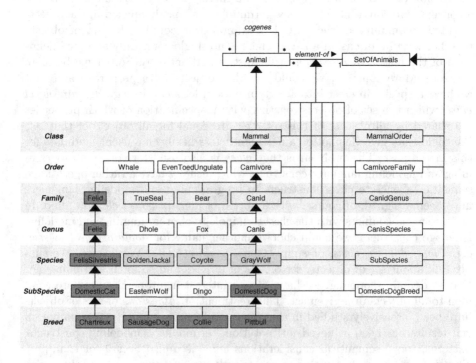

Fig. 6. The mammal hierarchy.

a concept of power set into the tier of class instances. Note, that even in set theory [90] all sets are objects. The class of sets, i.e., the set carrier, consists of opaque objects only. Inner structure of sets is an illusion that emerges in our minds by the application of the set forming operation $\{\_, \_\}$ on the carrier, i.e., the class of sets is an abstract data type. Sets of sets emerge by the axiom of pairing in the finite case and by the axiom of infinity in the infinite case. However, once a set of set is constructed you can also perceive it as a relationship between its members and itself, and equally, it is only a perception or illustration that membership means containment. And this is also true for predicate logic. Signatures in predicate logic can be considered pure, most possible reductionist entity-relationship models.

A credo often somehow stated in object-orientation is: *Everything is treated as an object.* In form-oriented analysis [26,37] we have expressed doubt in the metaphoric power of such and similar real-world statements. We have said that the value of such a statement is not clear if it is only used as a preamble or eye-catcher and is not exploited anywhere else in the subsequent methodology or its semantic foundation. Now, with the current discussion we have actually found a use case for this real-world statement. If it is a crucial value for object-orientation that everything is treated as an object, then also sets, sets of sets, sets of sets of sets and so forth should be treated as objects.

5.1   Plain Class Modeling for Sets of Sets of Domain Objects

Figure 7 shows a state of our *dogs and breeds* expert domain and a first simple yet adequate conceptual model for the intended domain. The diagrams (ii) through (iv) in Fig. 8 show further, more elaborate means to model the intended domain. Diagram (i) in Fig. 8 is, basically, the conceptual model copied from Fig. 7. It is included into Fig. 8 for an important presentation issue, i.e., in order to complete the full power type construction diamond.

The M0- and M1-level models in Fig. 7 together with the OCL_R constraints that are given in the sequel adequately represent the domain and the current state of the domain. The class Dog represents the set of dogs in the domain state. The set of dogs is a domain object that owns a genus, in this case Canis, as a property. All the dogs share the same property, therefore, this property is modeled as a class attribute in the M1-model, which is, as usual, indicated by underlining the attribute. Instead of assuming a singleton object as host for the class attribute, we explicitly specify this by the following OCL_R constraint:

$$
\begin{aligned}
&Dog.allInstances \rightarrow forAll(dog_1, dog_2 \mid \\
&\quad dog_1.genus = dog_2.genus \\
&\quad or\ dog_1.genus \rightarrow asSet \rightarrow size = 0 \\
&\quad or\ dog_2.genus \rightarrow asSet \rightarrow size = 0 \\
&)
\end{aligned}
\tag{46}
$$

We prefer to use the term class attribute over using the UML term static attribute. Because UML static attributes are not really static, but just class-global. A UML static attribute can vary over M0-model editions; however, what

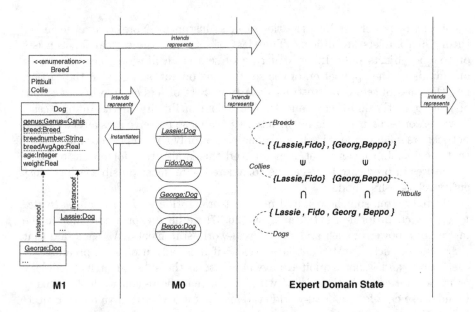

Fig. 7. Adequate constraint-based modeling of sets of sets without power types. The necessary OCL_R constraints are given in the text as (46)–(56)

is required for a UML static attribute is that it is equal for all objects of its hosting class in a given state. We have chosen the current formalization of the concept of class attribute, because it is amenable in a straightforward manner on basis of OCL.

A reader might say that constraint (46) is superfluous, because the UML specification states that a class attribute belongs to the class rather than to the objects of the class. However, the UML specification is not formal with respect to this, because it neither states the existence of a singleton object hosting the class attributes for each class nor does it mention class attributes in the semantic description of class instantiation. In that sense constraint (46) is one means to make the semantics of class attributes precise. However, the purpose of constraint (46) in this article is different, we want it to be at hand for comparison with the constraints for subset-global attributes like (48) to (51) in the sequel.

The constraint (46) is quite explicit and elaborate, it can be expressed much denser in a different style:

$$Dog.allInstances.genus \rightarrow asSet \rightarrow size \leqslant 1 \qquad (47)$$

Second, each dog has an age and a weight. These properties are, without loss of generality, different for each dog. Therefore, they are modeled as ordinary object attributes. Third, each dog has a breed, a breed number and the average age of its breed as a property. Again, these properties are different for each dog, but they are not completely arbitrary. Instead, there is a mutual functional dependency between the breeds and the breed numbers, and a functional dependency between breeds and average ages per breed. We choose the breed itself to identify the respective subset of dogs, i.e., collies or pitbulls. The enumeration type Breed hosts values Pitbull and Collie for this purpose. The properties breed number and breed's average age are not global with respect to the class dog but must be the same for all collies and independently the same for all pitbulls. We can express this by the following constraints:

$$Dog.allInstances \rightarrow select(breed = \#Collie).breednumber \rightarrow asSet \rightarrow size = 1 \quad (48)$$

$$Dog.allInstances \rightarrow select(breed = \#Collie).breedAvgAge \rightarrow asSet \rightarrow size = 1 \quad (49)$$

$$Dog.allInstances \rightarrow select(breed = \#Pitbull).breednumber \rightarrow asSet \rightarrow size = 1 \quad (50)$$

$$Dog.allInstances \rightarrow select(breed = \#Pitbull).breedAvgAge \rightarrow asSet \rightarrow size = 1 \quad (51)$$

Note, that the leading sign $\#$ in (48) through (51) is the usual way to denote enumeration literals in OCL. Furthermore, note that the constraints (48) through (51) must not be mixed with constraints of the following form:

$$Dog.allInstances \rightarrow select(breed = \#Collie) \rightarrow forAll(breednumber \rightarrow asSet \rightarrow size = 1)$$
$$Dog.allInstances \rightarrow select(breed = \#Collie) \rightarrow forAll(breedAvgAge \rightarrow asSet \rightarrow size = 1)$$
$$\dots$$

$$(52)$$

This means that the constraints in in (48) through (51) are not merely about multiplicities of properties as one might think at the first sight. Instead, they specify the uniqueness of the properties with respect to each breed. Multiplicities of properties are specified by the constraints of the form (52). They specify a [1..1] cardinality for the properties. In our example, the [0..1] cardinality is implicitly specified for the properties in diagram (i) in Fig. 7, because it can be assumed as the default cardinality of properties.

Later, when we consider subtype externalization in Sect. 5.3 we will discuss that these constraints can be expressed by turning the subset-global attributes into appropriate class attributes. The average age for collies can be the same as the average age of pit bulls. However, the breed number is regarded as the identifier in the domain. It must be different for collies and pitbulls. We can express this by the following constraints:

$$Dog.allInstances \rightarrow select(breed = \#Collie) \rightarrow forAll(c \mid$$
$$Dog.allInstances \rightarrow select(breed = \#Pitbull) \rightarrow forAll(p \mid$$
$$not(c.breednumber = p.breednumber) \quad (53)$$
$$)$$
$$)$$

Now, last but not least, let's have a look at the set of breeds in the expert domain. The set of breeds is represented by the enumeration type Breed at M1-level. The set of collies is represented by the set of M0-level objects that share the value Collie for their breed attribute. It is also represented at M1-level by the value Collie itself.

5.2 Making Constraints Robust Against M1-Level Model Updates

In Sect. 5.1 we have seen an important aspect of modeling sets of sets, i.e., subset-global attributes. We need to investigate these further and will introduce the notion of subclass attribute. Furthermore we need to discuss auxiliary properties for subsets of objects as well as properties that are global to sets of sets of objects. For this purpose, we investigate several options of modeling in Fig. 8. However, the central theme of this section turns out to be the question of how to make constraints robust against model updates at level M1.

The constraints (48) to (51) work fine to protect the M0-level objects against inadequate updates. However, in general, they are not sufficient for M1-level model updates. For example, if a new value, e.g., Beagle, is introduced by the modeler into the enumeration type, the subset-global attribute for breed numbers is *no longer under the auspices of appropriate constraints*, because the existing

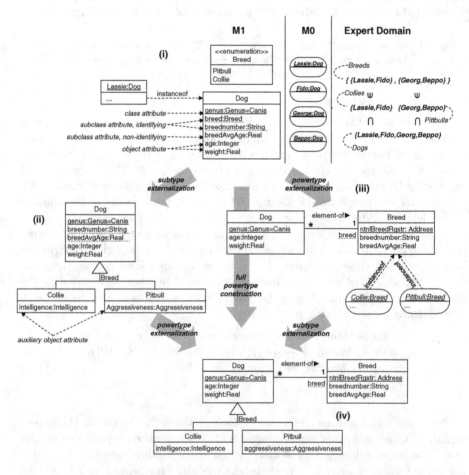

Fig. 8. Constraint-based modeling of sets of sets with UML modeling elements.

constraints work only for the former collection of enumeration type literals Collie and Pitbull, which infringes the intended meaning of the Breed enumeration type as representing the domain set of breeds.

Now, the following OCL_R constraints generalize the constraints (48) to (51) by abstracting from the concrete values in the enumeration type so that the constraints become robust against unwanted M1-level updates:

$$
\begin{aligned}
&\langle Breed \rangle \downarrow .ownedLiteral \to forAll(breedId \\
&\quad Dog.allInstances \\
&\quad \to select(breed = \langle breedId \rangle \uparrow).breednumber \to asSet \to size = 1 \\
&)
\end{aligned} \tag{54}
$$

$$
\begin{aligned}
&\langle Breed \rangle \downarrow .ownedLiteral \to forAll(breedId \\
&\quad Dog.allInstances \\
&\quad \to select(breed = \langle breedId \rangle \uparrow).breedAvgAge \to asSet \to size = 1 \\
&)
\end{aligned} \tag{55}
$$

See how, the constraints (54) and (55) make use of reification and reflection. The type *Breed* is a user defined type. It denotes an enumeration. Therefore, after reification of *Breed* we have access to its enumeration literals via introspective access. These can be, after the application of the reflection operator $\langle breedId \rangle \uparrow$, further exploited in M1-level constraint writing.

Let us call a constraint that is made robust against M1-level updates, a *sustainable constraint*. We will further delve into this terminology in Sect. 8. Next, we also want to turn constraint (53) into a sustainable version. This is even possible without OCL_R, i.e., in plain OCL. With constraints (48) and (50) we have specified that breed numbers are unique with respect to each breed. The fact can be exploited to give a less explicit and much more dense version of constraint (53):

$$
\begin{aligned}
&Dog.allInstances.breed \to asSet \to size \\
&= Dog.allInstances.breednumber \to asSet \to size
\end{aligned} \tag{56}
$$

As it turns out, the constraint (56) is independent of concrete breed identifiers and is therefore a robust version of (53).

Now, let us detour a bit and discuss the pragmatics of tool design. A more detailed discussion is provided by the symbolic viewpoints in Sect. 8. Assume that we have a tool that allows for modeling and instantiation of objects in parallel. Imagine that such a tool supports the maintenance of both the model and the data and, in particular, surveils the validity of reflective constraints, e.g., constraints written in OCL_R. Now, given such a tool, how to introduce a new breed into our example model? The answer is: (i) introduce a new value in the Breed enumeration type, (ii) instantiate some dog information objects, (iii) set the attributes of the new objects and care for the equality of the set-global attributes and the uniqueness of the breed number, (iv) submit the model changes as update and (v) expect the modeling tool to do the necessary constraint checking and reject resp. accept the changes based on the result.

5.3   Subtype Externalization

Model (ii) in Fig. 8 shows the result of externalizing the breed attribute into subclasses for each possible value. Having a certain value for an attribute, i.e., having a certain property, characterizes a subset of a set of objects, i.e., the set of objects sharing this property. Therefore, in model (ii) the subclasses Collie and Pitbull represent the domain sets of collies resp. pitbulls. The generalizations of the classes Collie and Pitbull to the class Dog form a UML generalization set which receives the name Breed in model (ii). This generalization set Breed now adequately represents the set of breeds in the expert domain.

Still, we need to enforce the global uniqueness of the breed number and average age with respect to the subsets of collies and pitbulls. We could get this effect by erasing the respective attributes from the class Dog and moving them as class attributes to the subclasses Collie and Pitbull. However, it is better OO-style to keep them in the class Dog so that they are inherited by the subclasses, for example, because we want to introduce further breeds as subclasses in future model editions. A means to override an attribute by a class attribute in a subclass is also no substitute for the given constraints, because without the constraints nothing ensures that the attribute is systematically overridden in all subclasses under consideration. In the current scenario, it is fair to call these attributes *subclass attributes*, because semantically they can be considered class attributes of the subclasses. We have therefore underlined them with a dashed line in the diagrams (i)–(iii) in Fig. 8. We do not want to introduce the concept of subclass attribute with this semantics as a language element here, because although it would work immediately for usual OO programming languages, it is incomplete in UML. In contrast to usual programming languages, generalization in UML can be non-disjoint [67], so in general you would also need to specify the subclasses for which the intended properties are considered as set-global.

The constraints (48) to (51) can now be re-stated for model (ii) as follows – note that the resulting constraints are actually class attribute constraints onto the considered attributes in their role as inherited attributes:

$$Collie.allInstances.breednumber \rightarrow asSet \rightarrow size = 1 \qquad (57)$$

$$Collie.allInstances.breedAvgAge \rightarrow asSet \rightarrow size = 1 \qquad (58)$$

$$Pitbull.allInstances.breednumber \rightarrow asSet \rightarrow size = 1 \qquad (59)$$

$$Pitbull.allInstances.breedAvgAge \rightarrow asSet \rightarrow size = 1 \qquad (60)$$

Now, in order to make the constraints (57) to (60) sustainable, i.e., robust against M1-level updates, we can restate the sustainable constraints (54) and (55), again in OCL_R, in terms of the generalization set *Breed*. The involved subclasses are the same with respect to their generalization set as the literals are with respect to their enumeration type:

$$\langle Breed \rangle \downarrow .generalization.specific \rightarrow forAll($$
$$\langle self \rangle \uparrow .allInstances.breednumber \rightarrow asSet \rightarrow size = 1 \qquad (61)$$
$$)$$

$$\langle Breed \rangle \downarrow .generalization.specific \rightarrow forAll($$
$$\langle self \rangle \uparrow .allInstances.breedAvgAge \rightarrow asSet \rightarrow size = 1 \qquad (62)$$
$$)$$

Next, we also give an equivalent to constraint (56):

$$\langle Breed \rangle \downarrow .generalization.specific \rightarrow asSet.size$$
$$= Dog.allInstances.breednumber \rightarrow asSet \rightarrow size \qquad (63)$$

With model (ii) the introduction of a new breed turns out to correspond to the introduction of a new subtype under the auspices of the necessary constraints. Model (ii) has an important advantage over model (i). The subclasses are the natural host for auxiliary attributes that are specific to a certain subset of domain objects. In the example we have chosen the attribute intelligence for collies and the attribute aggressiveness for pitbulls.

In principle it is possible to turn attributes of a each type into a subtype, i.e., not only attributes of an enumeration type. For enumeration types, which are finite, we simply turn each literal into a type, as we have seen in the current example. For an infinite type we update the model by the introduction of a new subtype representing a value of the type whenever necessary, i.e., whenever an attribute with this value occurs for the first time. This extreme subtype externalization is merely a thought experiment; but it is an instance of the purely symbolic viewpoint of modeling that we will discuss in Sect. 8.3, because it treats the evolving M1/M0-model as a single whole data store.

5.4 Power Type Externalization

Model (iii) shows the result of externalizing all the subset-global attributes in their own class Breed. It is usual to call a class like the class Breed a power type [46,47,68]. Now the concept of the set of breeds is made explicit by a class in the model. A concrete breed can now be represented by an M0-level object or an M1-level instance specification. This modeling solution might appear to the reader as particularly natural, because a class can be seen as the natural candidate to represent a set of objects, which are meant to be sets in this case. Actually, because of the 1-multiplicity at the element-of association, the constraints (48) to (51) become obsolete with solution (iii). Now, all we need to do is to generalize this situation to an arbitrary number of breeds is to adopt constraint (56) the following way:

$$Dog.allInstances.breed \rightarrow asSet \rightarrow size$$
$$= Dog.allInstances.breed.breednumber \rightarrow asSet \rightarrow size \qquad (64)$$

With the model (iii) there might be empty breeds due to the breed-to-dog association's many-cardinality [*]. If we want (63) to effect also empty breeds we need to change it to:

$$Breed.allInstances \rightarrow asSet \rightarrow size$$
$$= Breed.allInstances.breednumber \rightarrow asSet \rightarrow size \qquad (65)$$

Note, that both (64) and (65) are no OCL_R constraints, i.e., they are plain OCL constraints. With solution (iii) the membership of a dog in a breed is represented by instances of the element-of association. The model (iv) has an important advantage over the model (i). The class Breed is the natural host for auxiliary properties that are common to all breeds, e.g., the address of the national breed registry. Without the need for subtype-specific attribute extensions solution (iii) actually appears the most natural modeling pattern for the given scenario. This comes at no surprise: solution (iii) is no more, no less than the *type-object pattern* [52] of Johnson and Woolf. Unfortunately, we sometimes might want to model properties that are specific to certain breeds – see also the discussion on the disadvantages of implementation complexity in [52]. This leads us to the next Sect. 5.5.

5.5   Integrated Subtype and Power Type Externalization

Solution (iv) now shows an equivalent to the full UML power type construction [60,75] for the scenario. It makes explicit *(a)* the several breeds as subclasses Collie and Pitbull and *(b)* the set of all breeds as a class Breed. These two representations must now be balanced and kept in synch. First, we need an OCL_R constraint that expresses that all M0-objects of a given breed subclass are assigned to the same breed M0-object:

$$\langle Breed \rangle \downarrow .generalization.specific \rightarrow forAll(\\ \langle self \rangle \uparrow .allInstances.breed \rightarrow asSet \rightarrow size = 1 \\ ) \tag{66}$$

Second, we also need to express that objects of different subclasses are assigned to different Breed objects:

$$Dog.allInstances.breed \rightarrow asSet.size \\ = \langle Breed \rangle \downarrow .generalization.specific \rightarrow asSet \rightarrow size \tag{67}$$

The user-defined type name *Breed* is overloaded in diagram (iv). It denotes both the power type call *Breed* as well as the generalization class *Breed*. This does not pose a problem. In the constraints (66) and (67) the type *Breed* is used to denote the generalization set. Together, the constraints (66) and (67) imply that the subclasses of the *Breed* generalization set have no instances in common, i.e., that their sets of instances are disjoint. This is not automatically so. Multiple classification is, as a matter of course, an option with the UML, for example, because of multiple inheritance. UML generalization sets have an attribute *isDisjoint*, that specifies whether the specific classifiers of a generalization set may have instances in common or not – see [75]. We can specify that the subclasses of *Breed* are disjoint as follows:

$$\langle Breed \rangle \downarrow .isDisjoint \tag{68}$$

Note, that (66) and (67) together imply (68), but, however, the converse implication does not hold.

The two constraints (66) and (67) capture the essence of the UML power type construct, however, only for a special case. First, there must be exactly one power type association and, moreover, the involved power type association may be a many-to-one association only, see the element-of association in Fig. 8 in our case. In Sect. 7 we will provide general constraints for arbitrary user-defined power type specifications.

6   Z and Sustainable Constraint Writing

In this section we restate the domain model from Fig. 7 in the specification language Z. The aim of this is twofold. First, the specification offers a particularly dense presentation of the crucial domain knowledge discussed throughout the article and is amenable to foster its understanding. In that sense, we will refer to this Z example later in Sect. 7 on the precise semantics of UML power types. Second, and maybe even more important, its discussion can foster the understanding that semantics and pragmatics are concerns in language design that can and should be separated – and this is so also, and in particular, in case of modeling languages.

The specification language Z allows describing system states on the basis of set theory and predicate logic. It offers rich notation for all usual mathematical constructs. It is an advantage to have a standardized means to write mathematical specification. However, Z is more than a neat set notation. It establishes a system model and a system modeling paradigm. A system is modeled as a state evolvement. The approach is to model the state transition as manipulation of declared functions (pre-post-condition specification). It belongs to the large family of Parnas methods [78] with ASMs (abstract state machines) as a most recent member [40]. We use only the data facet of Z in this article. I recommend [87] as a reference, and also [43,44,61]. Furthermore, the Z notation is standardized by an ISO standard [49].

Z specifications are not automatically sustainable. However, they can be turned into sustainable specifications. No reflective refactoring of Z is needed for this purpose, because Z allows for quantification over arbitrarily nested sets. It is common, e.g., in text books on Z, to say that the Z notation is a combination of set theory and first-order logic. But take care; this is not a formal statement. Informally, it is a neat explanation. Formally, it can neither be neglected nor approved, because it is not clear what is meant by *combining* set theory and first-order logic. In any case, it is important to understand, that in Z it is possible to quantify over arbitrarily nested sets. So, if Z had a sufficiently formal semantics, it would be in the realm of a typed, higher-order logics, comparable to Isabelle/HOL [64], see also [54,83] for a discussion. The way we turn a Z specification of our example domain into a sustainable Z specification in Sect. 6.3 is very instructive and gives us yet another viewpoint onto today's object-constraint languages, their expressive power and their pragmatics.

6.1 Types Specification

We introduce the basic sets of dogs, addresses, genera, breed numbers, intelligence degrees and aggressiveness degrees. There are all kept completely opaque in the following:

$$[DOG]$$
$$[ADDRESS, GENUS] \qquad\qquad (69)$$
$$[BREEDNUMBER, INTELLIGENCE, AGRESSIVENESS]$$

It would be typical Z style to introduce further, derived types for intended domain concepts. For example it would be typical to introduce the following type for breeds:

$$BREED == \mathbb{P}\, DOG \qquad\qquad (70)$$

There are two good reasons for auxiliary types as (70). First, they can improve the self-documentation of the specification. Second, they improve re-use. It is typical Z style to make intensive use of such auxiliary types. However, in our case, we stay with the plain types given in (69), because this eases the discussion. Our interest in this section is the discussion of design principles, whereas the artifact quality of the specifications play a minor role.

6.2 Schema Specification

The structure of possible system states manifests in variables and axioms declared in schemas. Mathematical notation is the first class-citizen in Z. For the modeler this means, that he must often specify concepts that would be available as syntactic sugar in other modeling languages. Nevertheless, Z specifications are usually rather dense than bloated. The advantage is that we can hardly deviate from the declarative, mathematical semantics. With the schema *DogDomainData* in (71) we provide a straightforward specification of the system state. With the schema *DogDomainAttributes* in (72) we add the attributes – compare this to the domain model provided in Fig. 7. In our Z specification we model each attribute as a function that yields a value for each given parameter object. This means, we explicitly model an object-mechanism that is implicitly given in each object-oriented modeling language – see also the ephemeral object patterns in [3, 4] for a discussion of this specification style.

Each variable in (71) represents a part of the system state. Therefore each variable holds a subset of its corresponding base type and is typed as power set of this. The set *dogs* stands for the set of dogs at one point in time, whereas the type *DOG* stands for the set of all possible dogs that may ever exist in any of the system states. Consequently the type of the set *dogs* is modeled as the power set of *DOG*. Similarly, the type of the set *breeds* is modeled as the power power set of *DOG*. At each point in time the set *breeds* consists of sets of dogs. Furthermore, we specify that both of the two breeds *collies* and *pitbulls* are subsets of the set of dogs in each system state. Furthermore, we specify that the two breeds *collies* and *pitbulls* are always disjoint sets.

$$
\begin{array}{|l r|}
\hline
\_DogDomainData\_ \\
dogs : \mathbb{P}\,DOG \\
breeds : \mathbb{P}\,\mathbb{P}\,DOG \\
collies : \mathbb{P}\,DOG \\
pitbulls : \mathbb{P}\,DOG \\
\hline
collies \subseteq dogs & (i) \\
pitbulls \subseteq dogs & (ii) \\
collies \in breeds & (iii) \\
pitbulls \in breeds & (iv) \\
collies \cap pitbulls = \emptyset & (v) \\
\hline
\end{array}
$$

$$(71)$$

The attributes of the classes of our example are modeled as partial function from the sets of potential objects, i.e., types, into their value ranges in schema (72). A major role of the schema (72) is then, to specify the correct domains of the attribute functions. There is no need to specify the uniqueness for breed numbers and breed average ages for the members of a given breed in the Z solution. This is so, because the corresponding attributes are modeled as functions that have the set of breeds as their domain. The functions assign values to breeds, not to dogs. This solution corresponds to the power type externalization solution in Sect. 5.4, in which these attributes were modeled as properties of power type objects.

$$
\begin{array}{|l|}
\hline
\_DogDomainAttributes\_ \\
DogDomainER \\
dogGenus : GENUS \\
dogAge : DOG \nrightarrow \mathbb{N}_0 \\
dogWeight : DOG \nrightarrow \mathbb{R} \\
breedNationalBrgRegistrar : ADDRESS \\
breedNumber : \mathbb{P}\,DOG \nrightarrow BREEDNUMBER \\
breedAvgAge : \mathbb{P}\,DOG \nrightarrow \mathbb{R} \\
collieIntelligence : DOG \nrightarrow INTELLIGENCE \\
pitbullAgressiveness : DOG \nrightarrow AGRESSIVENESS \\
\hline
\mathrm{dom}\ dogAge = dogs \\
\mathrm{dom}\ dogWeight = dogs \\
\mathrm{dom}\ breedNumber = breeds \\
\mathrm{dom}\ breedAvgAge = breeds \\
\mathrm{dom}\ collieIntelligence = collies \\
\mathrm{dom}\ pitbullAgressiveness = pitbulls \\
\hline
\end{array}
$$

$$(72)$$

Now, let us compare the Z solution to the power type solution in Sect. 5.5. Constraints (i) and (ii) in the schema *DogDomainData* correspond the introduction of *Collies* and *Pitbulls* as subclasses in the generalization set *Breed* in Sect. 5.5. In the power type solution we had to balance the subclasses of the full power type construction with the power type objects. It is constraint (66) that enforces a unique power type object for all objects of a given subclass. The constraints (iii) and (iv) for *collies* and *pitbulls* in the schema *DogDomainData* can be considered the counterpart of constraint (66) in Sect. 5.5. The constraints (iii) and (iv) are particularly simple. Let us have a look at the set *collies*. The set *collies* is itself both a subset of the set *dogs* and at the same time an element of the set *breeds*. This is possible, because Z has the expressive power of a typed, higher-order logic. There is no need for an extra object representing the set *collies* as a whole. Attributes that are common to all collies are simply assigned to the whole set *collies*. All this is also true for the set of *pitbulls*. Therefore, there is no need to balance the set of *collies* and *pitbulls* against objects that represent.

Next, the constraint (67) in Sect. 5.5 enforces that the set of instances of *Collies* is disjoint from the set of instances of *Pitbulls*. Therefore, constraint has (67) constraint (v) in the schema *DogDomainData* as its counterpart in the Z solution. The Z constraint (v) is again particularly easy due to the fact that we can exploit mathematical set notation for it.

6.3 Sustainable Schema Specification

The crucial constraints (i), (ii) and (iv) in schema (71) are not sustainable. If we add a new breed, let's say *beagles* $\in$ *breeds*, it is neither ensured, that the new breed is a sub set of the set of dogs, nor that it is disjoint to the already existing breeds. Let us have a look at schema (73) which is a solution to this problem. Constraint (i) in (73) is an appropriate sustainable generalization of the constraints (i) and (ii) in (71). Constraint (v) in (73) is an appropriate sustainable generalization of its counterpart (v) in in (71).

$$
\begin{array}{l}
\underline{\quad DogDomainDataSustainable \quad\quad\quad\quad\quad\quad\quad\quad} \\
dogs : \mathbb{P}\, DOG \\
breeds : \mathbb{P}\,\mathbb{P}\, DOG \\
collies : \mathbb{P}\, DOG \\
pitbulls : \mathbb{P}\, DOG \\
beagles : \mathbb{P}\, DOG \\
\hline
\end{array}
$$

$$
\begin{array}{ll}
\forall\, breed : breeds \bullet breed \subseteq dogs & (i) \\
collies \in breeds & (ii) \\
pitbulls \in breeds & (iii) \\
beagles \in breeds & (iv) \\
\forall\, breed_1, breed_2 : breeds \bullet breed_1 \cap breed_2 = \emptyset & (iv)
\end{array}
$$

$$(73)$$

It also would have been possible to add explicit constraints for the new beagle breed to the schema *DogDomainData*, i.e.:

$$beagles \subseteq dogs \tag{74}$$
$$beagles \cap collies = \emptyset \tag{75}$$
$$beagles \cap pitbulls = \emptyset \tag{76}$$

But, of course, the solution of schema (73) is better. Explicit constraints are again non-sustainable. They are not generic and therefore explicit constraints do not scale. Already in the current small example, they start to bloat the specification.

7   Precise Semantics of UML Power Types

The current UML superstructure specification contains the following description of the semantics of power types [75]:

> *Formally, a power type is a classifier whose instances are also subclasses of another classifier. [...] As established above, the instances of Classifiers can also be Classifiers. This is the stuff that meta models are made of.*

The statement is inconsistent against the background of the rest of the UML specification [75]: an M1-level subclass is an instance of the M2-level class *Class* and cannot be an instance of an M1-level classifier. Instances of an M1-level classifier cannot be classifiers themselves. Instances of an M1-level classifier are M0-level model elements and definitely do not reside at level M1. We must not mix the level-crossing UML instantiation relation with the set membership relation $\in$ in the intended domain. If you model with a power type construct the resulting model is not *per se* a meta model. Furthermore, the above statement is not a formal statement, but this is actually a minor point.

Where does the confusion stem from? One source of misunderstanding of the domain-relation $\in$ as level-crossing instantiation may arise from using the phrase *is instance of* for *is element of* in the domain, which might be natural in many domains. Compare this to the Z specification of the running example in Sect. 6. In (71) *collies* is an element of *breeds* and a the same time a *subset* of *dogs*. We must not mix modeling with the linguistic modeling framework that we exploit as tool, i.e., we must not mix $\in$ with the instantiation of a sentence of our modeling language which is described by its grammar, i.e., a meta model. We should never forget that meta models are really just kinds of grammars and we should not be confused by the fact that we use a common modeling language as notation and mechanism to write these grammars.

However, it is not appropriate to simply reject the above statement from the UML specification and similar statements in the community as inconsistent. It implicitly contains an important aspect of power types that goes beyond their meaning as constraining states of information objects of the current model. Based on the findings and terminology of this article, we can attempt an informal, yet more precise re-formulation of the above definition. For example, we could state:

Informally, a power type is a class whose instances represent sets of domain objects, where each of these domain objects is represented by an instance of a subclass of another classifier.

Arbitrary subclasses? An arbitrary classifier? No. All the extra information expressed by the constraints (66) and (67) is yet still missing, so that both the UML definition of power type as well as are our re-formulation yield no complete specification. Again compare the above statements with the Z specification in (71).

Now, with OCL_R we can give a general semantics for UML power types. With UML, a concrete user-defined power type consists of a generalization set and a designated power type class for this generalization set. The UML specifies that a generalization set has a property *powertype* of type *Classifier*, which is optional, i.e., has [0..1] cardinality. Obviously, the generalization set specifies a power type construct, whenever this value is present. Concrete power type objects can be assigned to the objects of the generalization set. Our specification needs to decide upon pramatic issues, i.e., we assume that there are associations between the super class of the generalization set and the power type class in order to assign concrete power type objects, which is in accordance to the literature and the examples in the UML specification. So far, in Sects. 5 and 6 we have treated examples of a special case, in which there is exactly one such association, which has to be, furthermore a many-to-one association. This special case is the usual case, e.g., all of the examples in the UML specification follow this pattern – see Figs. 7-49, 7-50 and 7-51 in [75]. We treat a most general case here – see Fig. 9. There might be many power type associations of arbitrary cardinalities and, furthermore, the generalization set is not yet required any more to be disjoint.

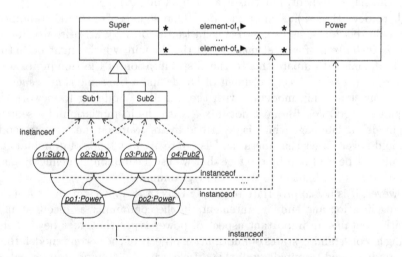

Fig. 9. General UML powertype specification pattern.

In the sequel we call an object of one of the subclasses of the generalization set a *generalization set object* for short. Furthermore, we will call a power type object that is assigned to a generalization set object a *representing object*. Now, we need to specify that (i) there is at least one representing object for each generalization set object, (ii) the number of representing objects equals the number of generalization set subclasses, (iii) for each representing object there exists a generalization set subclass, so that the given representing object is assigned to all objects of this subclass. This is achieved by the following OCL_R constraint:

```
01  ⟨Class⟩↓ .allInstances.forAll(Super, Subs, Power |
02    ⟨GeneralizationSet⟩↓ .allInstances → exists(gs |
03      gs.powertpye = Power
04      and   gs.generalization.general → includes(Super)
05      and   gs.generalization.specific = Subs
06    )
07    implies(
08      let ps = Super.ownedAttribute → select(type = Power) in (
09        ⟨Super⟩↑ .allInstances → forAll(o |
10          o.⟨ps⟩↑→ asSet → size ⩾ 1
11        )
12      and
13        ⟨Super⟩↑ .allInstances.⟨ps⟩↑→ asSet → size
14        = Subs → asSet → size
15      and
16        ⟨Super⟩↑ .allInstances.⟨ps⟩↑→ forAll(op |
17          Subs → exists(Sub |
18            ⟨Sub⟩↑ .allInstances → forAll(o |
19              o.⟨ps⟩↑→ includes(po)
20            )
21          )
22        )
23      )
24    )
25  )
```
$$(77)$$

Lines 02 through 06 establish all triples of classes *Super*, *Subs* and *Power* that correspond to a valid and complete user-defined UML power type. Here *Super* stands for *superclass* and means the general class of the generalization set, *Subs* stands for *subclasses* and means the collection of specific classes of the generalization set and *Power* stands for the *powertype* that is assigned to the generalization set – see Fig. 9 once more. Now, the sub constraint in lines 09 through 11 ensures property (i) from above, the sub constraint in lines 13 and 14 ensures property (i) and the sub constraint in lines 16 through 22 ensures property (iii). Altogether, constraint (77) grasps the essential semantics of UML power types. Each sub class is represented by a power type object. A power type object carries information that is common to all objects of the subclass it represents.

Still, there remain some pragmatic issues that remain open. In case of a complete overlap of the instances of two subclasses of the generalization set, we cannot distinguish between the two involved power type objects representing the subclasses any more. We do not design solutions to issues like that here. Instead, we turn to the special use case of UML power types, in which (i) there exists exactly one power type association, which (ii) then has a many-to-one cardinality. In this case, we need to specify, that (iii) for each generalization set subclass the same representing object is assigned to all objects of the given subclass and (iv) different representing objects are assigned to the objects of different generalization set subclasses. This means that we need to generalize constraints (66) and (67) from Sect. 5.5 to all user-defined power types. This is achieved by the following OCL_R constraint:

$$
\begin{aligned}
&01 \ \langle Class \rangle \!\downarrow .allInstances.forAll(Super, Subs, Power \mid \\
&02 \quad \langle GeneralizationSet \rangle \!\downarrow .allInstances \rightarrow exists(gs \mid \\
&03 \qquad gs.powertpye = Power \\
&04 \qquad and \quad gs.generalization.general \rightarrow includes(Super) \\
&05 \qquad and \quad gs.generalization.specific = Subs \\
&06 \quad) \\
&07 \quad implies(\\
&08 \qquad let \ p = Super.ownedAttribute \rightarrow select(type = Power) \ in \ (\\
&09 \qquad\quad p \rightarrow asSet \rightarrow size = 1 \\
&10 \qquad\quad and \\
&11 \qquad\quad \langle Super \rangle \!\uparrow .allInstances \rightarrow forAll(o \mid \\
&12 \qquad\qquad o.\langle p \rangle \!\uparrow \rightarrow asSet \rightarrow size = 1 \\
&13 \qquad\quad) \\
&14 \qquad\quad and \\
&15 \qquad\quad Subs \rightarrow forall(Sub \mid \\
&16 \qquad\qquad \langle Sub \rangle \!\uparrow .allInstances.\langle p \rangle \!\uparrow \rightarrow asSet \rightarrow size = 1 \\
&17 \qquad\quad) \\
&18 \qquad\quad and \\
&19 \qquad\quad \langle Super \rangle \!\uparrow .allInstances.\langle p \rangle \!\uparrow \rightarrow asSet \rightarrow size \\
&20 \qquad\quad = Subs \rightarrow asSet \rightarrow size \\
&21 \qquad) \\
&22 \quad) \\
&23 \ )
\end{aligned}
\tag{78}
$$

The sub constraint in line 08 ensures property (i) from above, the sub constraint in lines 11 through 13 ensures property (ii), the sub constraint in lines 15 through 17 ensures property (iii) and the sub constraint in lines 19 and 20 ensures property (iv).

8   A Symbolic Viewpoint of Modeling Languages

Figure 10 shows different viewpoints on model evolution. Note, that they are really only viewpoints on one and the same scenario. Each of the viewpoints

grasps important issues in pragmatics of information system design and operations. Furthermore, the distinction between ephemeral versus evolution persistent constraint writing in Fig. 10 is an important concept in its own right.

8.1   The Classic Database Evolution Viewpoint

The first viewpoint (i) in Fig. 10 is the classic database viewpoint, which is also
the usual OO programming language viewpoint. The schema is given as an OO
class diagram and is cleanly separated from the data. The schema corresponds
to the UML M1-level, whereas the data corresponds to the UML M0-level. It
is assumed that the schema is fixed, whereas the data is not. The data is continuously manipulated. This viewpoint therefore distinguishes between design
time and runtime. The schema shapes the information space. It constraints the
structure in which we can capture and maintain data. However, it is also possible to fix more complex domain-related integrity constraints for the data, for
example, referential integrity, class-internal functional dependencies, or domain-
related integrity constraints, e.g., the rule that a certain integer value must not
exceed a maximum value and so forth. A crucial feature of databases is to support the enforcement of these constraints that are considered an integral part of
the schema. Whenever you try to update the data in a way that would violate
the constraints, the database will reject your update.

   We have said that the schema is fixed. But actually it is not. Schema updates
can occur. However, it is important to understand that in the viewpoint
(i) schema updates are considered to occur seldom and therefore schema updates
are considered almost fixed. *Seldom* and *almost* are vague concepts and therefore we will be able to switch to the equal M1/M0 resp. symbolic model evolution viewpoint (ii) later. Furthermore, schema updates are regarded as cost-
intensive and are usually controlled by other access rights than those for data
updates. Usually, you need to contact your database administrator for this purpose. Whenever a schema update occurs, it triggers a data migration step as
indicated by the numbers *1* and *2* in Fig. 10. This data migration step can be
very complex, because the existing data must be re-shaped [14,27,28,32,53].
Similarly, if you change the class structure of your application this at least
means that you need to stop, recompile and restart the application program.
For an enterprise application this can already be very cost-intensive and risky.
Hopefully, the program has been designed for reuse and the change has been
foreseen in the applied patterns. If not, and if your changes are really structural,
unforeseen changes, this can easily give rise to a cost-intensive code refactoring
project.

8.2   The Symbolic Viewpoint

The classic database viewpoint is pervasive. For example, the UML meta level
architecture distinguishes between an M1-level and M0-level – note, that the
M0-level is explicitly called the runtime object level in the UML specification.
Nevertheless, the viewpoint is not set in stone. It is simply possible to view

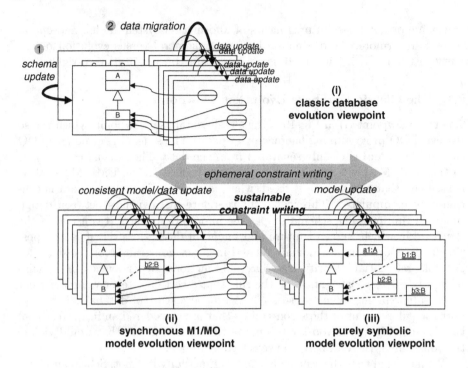

Fig. 10. Viewpoints on model evolution.

schema and data updates as equal. Once we abstract from the differences in frequency, costs and access rights for schema and data updates, the way is free to review the scenario from a different light. First, the M1-level model elements also encapsulate information about the *intended* [15,16] domain, not only the M0-level objects! In that sense, the M1-level is also a data level. Second, there can be also important constraints on the M1-level model elements with respect to the domain. These can be completely independent from the M0-level. And more importantly, it might be adequate to state them in terms of potential, i.e., not yet instantiated M1-level model elements. For example, you might have a class hierarchy that consists of two trees and might want to ensure that whenever a new subclass is added to one of the trees, a further subclass should also be added at the same position into the other tree. Usual database technology will not support the application of such constraints when updating schemas.

We call a constraint that is written in terms of only a fixed number of concrete M1-level model elements an *ephemeral constraint*, if it may fail to fulfill its intended purpose after an update of some M1-level model elements. Obviously, this description of ephemeral constraints is not a strict definition and the notion of ephemeral constraint is therefore an informal notion. A constraint is ephemeral only with respect to a certain notion of considered model update and a certain notion of intended purpose. We call a constraint that overcomes the weakness of an ephemeral constraint an M1-level model evolution persistent constraint, evolution persistent constraint, *persistent constraint* or *sustainable constraint* for short. Usually, a sustainable constraint is achieved by generalizing an ephemeral constraint to all possible relevant user-defined types.

Once we have adopted an equal M1/M0 model evolution viewpoint, we are free to think about new tools with innovative modeling features. However, we must not forget about the established database viewpoint, because it incorporates the important aspects of cost-effects and access right management. Furthermore, viewpoint (ii) enables us to rethink the semantics of modeling elements in order to make it more precise.

We call viewpoint (ii) a *symbolic viewpoint*, because it stresses the fact that M1- and M0-level modeling elements can be considered as together intending [15, 16] objects in the expert domain. In terms of symbolic computation the M0-level modeling objects can be regarded as ground terms. For example in the UML, this viewpoint is obfuscated by the existence of instance specifications at level M1, in particular, because instance specifications are optional. Therefore, we introduce the purely symbolic viewpoint (iii) in Fig. 10 as a refinement of viewpoint (ii).

8.3 The Purely Symbolic Viewpoint

In the purely *symbolic viewpoint* (iii) we assume that all M0-level objects are always and only captured and maintained by instance specifications that represent them. For example, as a thought-experiment, we could design a database based on UML class diagrams in which we capture and manipulate database objects always and merely by instance specifications.

The purely symbolic viewpoint can help to avoid certain confusions. In the discussion of OO semantics it can easily happen – and happened in the past – that distinct concepts like the following are thrown together and confused with each other: *(a)* instantiations of M1-level elements, *(b)* instances of M1-level elements, *(b)* set memberships in the expert domain, *(c)* representations of instances of M1-level elements at level M1, *(d)* instantiations of M2-level elements, *(e)* instances of M2-level elements, *(f)* representations of set memberships in the expert domain at level M0, *(g)* representations of set memberships in the expert domain at level M1, *(h)* types at level M1, *(i)* classes at level M1, *(j)* class constructs of modeling languages, *(j)* intensions of sets of domain objects, *(k)* domain objects, *(l)* the intended meaning of domain objects, and so on and so forth.

You can perceive the achieved M0-level free modeling in two ways. Either practically, as a concrete tool in which the visual modeling canvas is also the data store, or simply as an appropriate formal viewpoint. Because, even if we discuss without M0-level, tools and languages can provide different interfaces or look&feels for the manipulation of the ground terms and the type terms. If we assume that all data is kept and maintained at M1-level this greatly eases and unifies the discussion. Note that in symbolic computation there is also no dedicated grammatical tier for the ground terms. In the reductionist calculi of symbolic computation like the lambda calculus [11] or PCF (Programming With Computable Functions) [80] the objects resulting from computations are terms of ground type, but still, they are just terms of the language and so it is the same with full-fledged functional programming languages or term rewriting systems. The symbolic viewpoint is a grammatical viewpoint. In a symbolic viewpoint, every object of interest is symbolized as term of the same language. This is the reason, why have chosen to call the viewpoint discussed here a symbolic viewpoint.

8.4   UML Instance Specification

Let us analyze UML instance specifications from the purely symbolic perspective of Sect. 8.3. Instance specifications represent M0-level objects. Let us have a look at our tiny example model in Fig. 7. Here we have M0-level objects *Lassie : Dog*, *Fido : Dog*, *George : Dog* and *Beppo : Dog*. Two of them, i.e., *Lassie : Dog* and *George : Dog* have also a UML-instance specification at M1-level. The UML considers instance specifications as examples only. There is explicitly no need to give an instance specification for each M0-level instance. Furthermore, an instance specification needs not to provide a slot and value specification for each attribute of the corresponding object. However, if we visualize an M0-object by an instance specification at level M1 it would make sense to require that an instance specification should obey to the same rules that we impose as constraints for the M0-level objects. We can do this with appropriate reflective constraints. Let us have a look at a first example, i.e., at the very basic constraint (1):

$$\textbf{context } Person \textbf{ inv: } age \geqslant 40 \tag{79}$$

We can turn (79) into an OCL_R constraint that appropriately effects instance specifications the following way:

$$
\begin{aligned}
&\langle InstanceSpecification \rangle \downarrow .allInstances \\
&\rightarrow select(classifier \rightarrow includes(\langle Person \rangle \downarrow)) \\
&\rightarrow select(slot \rightarrow forAll(\\
&\qquad definingFeature.name = \texttt{"}age\texttt{"} \\
&\qquad implies \\
&\qquad value.IntegerValue() \geqslant 40 \\
&\quad) \\
&)
\end{aligned}
\tag{80}
$$

Now, let us consider a constraint that also contains a navigation expression. Assume that we also have a class *Dog* with property *owner : Person*[0..*]. Now, consider the following constraint:

$$\textbf{context } Dog \textbf{ inv: } owner.age \geqslant 40 \tag{81}$$

Again, we can turn constraint (81) into an appropriate OCL_R constraint for M1-level instance specifications:

$$
\begin{aligned}
&\langle InstanceSpecification\rangle \downarrow .allInstances \\
&\rightarrow select(classifier \rightarrow includes(\langle Person\rangle \downarrow)) \\
&\rightarrow select(person \mid \\
&\quad \langle InstanceSpecification\rangle \downarrow .allInstances \\
&\quad \rightarrow select(classifier \rightarrow includes(\langle Dog\rangle \downarrow)) \\
&\quad \rightarrow exists(slot \rightarrow includes(\\
&\qquad definingFeature.name = "owner" \\
&\qquad and \\
&\qquad value \rightarrow includes(person) \\
&)\ )\ ) \\
&\rightarrow select(slot \rightarrow forAll(\\
&\quad definingFeature.name = "age" \\
&\quad implies \\
&\quad value.IntegerValue() \geqslant 40 \\
&)\ )
\end{aligned}
$$

(82)

Note, once more, that the UML allows instance specifications to be partial specifications, i.e., an instance specification does not have to specify a value for each property of the object that is represents. This explains the usage of *implies* in constraint (80). Constraint (80) allows for instances specifications that do not have a slot for the property *age*, however, if such a slot exists, it has to adhere to the given constraints. It is possible to change exactly this partial specification approach. It is possible to give OCL_R constraints that enforce that each instance specification is a full-fledged, consistent object description. Furthermore, the purpose of constraints (80) and (82) has been to demonstrate, that it is, in principle, possible to turn each OCL constraint into an appropriate OCL_R constraint on instance specifications. We do not give the detailed specifications of all this here.

The OCL_R constraints resulting from the described transformation are substantially more complex than the original constraints. Therefore, you might want to think of all the discussion here as a mere thought-experiment. However, it shows that we could get rid of the M0-level to achieve a purely symbolic viewpoint. It is important to understand that the M1- and M0-level together form a language to describe states in the expert domain. The existence of instance specifications merely introduces redundancy. Currently the semantics of UML relies on the notion of M0-objects, and, even more important, its constraint language OCL is designed in terms of M0-objects. We guess that the intention of M0-objects in the UML was to deliberately introduce a degree of freedom in the interpretation of models, i.e., in the sense that M0-objects could be, e.g., data objects in a database, or, run-time objects of an object-oriented programming language and so forth. Merely throwing away the M0-level, without appropriate tool support, is not really an option. However, the discussion also shows that it is actually possible to design appropriate tools for supporting a symbolic modeling viewpoint.

9 Related Work

Programming languages and their type systems, in particular, generative programming languages [20], form a mature field of study that is important for the current discussion. In the programming languages GENOUPE [33,34,57] and FACTORY [25,35] it is possible to implement OCL_R constraints. GENOUPE is a C#-extension, whereas FACTORY is a Java-extension. With GENOUPE and FACTORY generators it is possible to analyze a given class and weave its attributes as class attributes into another class. With these generators the languages are expressive as *DeepJava* [55]. *DeepJava* offers neat clabject-style syntax. The natural candidate for representing sets of sets in C# and Java is, the nested resp. inner classes construct [39]. The problems with nested classes is that subclassing cannot crosscut the nesting structure, which makes impossible a direct, natural transformation of, e.g., model (v) in Fig. 8 into code. This problem is even not overcome by nested inheritance as provide by *Jx* [65] and *J&* [66] or advanced nested composition constructs as provided by DEEPFJIG [19].

Generative programming can be understood in a very concrete, narrow sense. Then, it is about programming languages that offer generative programming language features and establish appropriate type systems for generative programming. Actually, the systematic generation of parts of software systems, in first place code, but also all other kinds of software artifacts, is a practically highly relevant topic and actually a widespread issue in professional projects. It comes along in many faces and flavors: domain-specific languages [22], compiler-compilers [77], rapid-development tools, object-relational mapping tools [14,28, 32], object-oriented component technologies, enterprise computing frameworks and so on and so forth. The Adaptive Object Model Architecture of Yoder and Johnson [81,89] is a mature approach to describe self-referential, systematically adoptable software systems, as well as their design patterns and architectural patterns.

In [13] the authors define the conceptual programming language PCF_{DP} as an extension of PCF (Programming with Computable Function) by a quotation mechanism known from LISP that allows for reification and reflection. PCF [80] is the typed lambda-calculus with recursion and can be considered a reductionist functional programming language. Then, the authors give an axiomatic semantics [42] for PCF_{DP} and this way achieve a program logics for generative run-time meta-programming.

Clabject modeling [6,8] is the established and major multilevel modeling approach [7]. With clabject modeling useful terminology has been created for the distinction between the different kinds of instantiations. In [9] the authors distinguish between so-called linguistic and ontological instantiation. Linguistic instantiation stands for the model-level-crossing instantiation relation. Ontological instantiation stands for the domain-level-crossing relation $\in$ in the intended domain. Clabjects are classes that allow for deep instantiation. The ontological instance of a clabject is itself a clabject and can therefore stand for a set of objects. This way it is possible to adequately represent arbitrarily nested sets, i.e., the syntactical rules of the clabject frameworks are suitable to guarantee the intended meaning of the model.

In [56] it has been clarified that meta levels must not be confused with the levels of a modeling hierarchy and also, that linguistic instantiation must not be confused with ontological instantiation. In [82] the authors elaborate a formal semantics, based on category theory [12], for terminology that has been created in the multilevel modeling community. *Nivel* [2] is a reductionist multilevel modeling language that supports clabjects, associations, generalization sets, but no power types. A formal description of *Nivel* is provided by translation to the Weight Constraint Language [86], i.e., stable model semantics. This formal description achieves a reformulation of the clabjects rules [10] in type systems notation [17].

Meta modeling tools are the natural candidates for supporting multi-level modeling and clabject modeling. They are also the natural, potential host for pervasive M2/M1/M0-level crossing constraint checking features. The tool MELANiE [6,8] already offers a clabject-oriented constraint language for this purpose. With an appropriate clabject modeling tool like MELANiE [6,8] we can assume that all information is represented at M1-level without any M0-level objects, i.e., without linguistic instances of classes. Therefore, clabject modeling tools also establish a modeling viewpoint that is similar to the purely symbolic viewpoint developed in Sect. 8.3. METADEPTH [23,24] is an implementation of a multilevel modeling language on the basis of the AToM^3 [88] meta modeling tool [88]. It supports clabjects as crucial concept and also checks for adherence to the clabject rules.

The meta modeling tool AMMI [30,31,48] defines and realizes the so called *visual reification principle*. Visual reification must not be mixed with the reification operators discussed for the OCL$_R$ semantics here. Visual reification is a kind of bipartite instantiation principle in meta modeling tools that allows for making meta models visually reminiscent of their own instances, which eases meta modeling for domain experts.

The symbolic viewpoints from Sect. 8.3 superficially resemble but must not be confused with the viewpoint of the important strand of research on *models and evolution* [21,84]. The *models and evolution* viewpoint incorporates potentially many kinds of artifacts with models as centrally important artifacts. It deals with the gaps between these artifact groups. It is a particularly mature but still classical viewpoint. Our symbolic viewpoints deal with models only and deliberately abstract from differences between different kinds of models. Our symbolic viewpoints are merely instructive devices that gain their value only from their tension with classical viewpoints on modeling.

Investigations on the relationship of OO conceptual modeling and ontological modeling are very promising [45,62] and have impact [50] – see [63] for a sound overview. For the understanding of the arguments in this article, the established mainstream interpretation of OO conceptual modeling as an extended, mature semantic modeling approach [18] is sufficient. In form-oriented analysis [26,36,37] we have characterized conceptual modeling as the school of shaping and maintaining information. We have identified real-world metaphors as being merely guidelines for requirement elicitation. This means, that for the argumentation in this article it is not necessary to understand conceptual modeling in terms of ontological modeling [38], i.e., as construction of an ontological commitment as characterized in [41].

We believe that the viewpoint of *considering models as evolving data storing systems* is also particularly appropriate for the emerging paradigm of cloud-based software engineering [59], which eventually demands, in our opinion, for a more holistic approach to the design of data services and their utilization [3,4]. For example, in [5] we have coined the concept of *viable software system* which is about systems that are pro-actively designed, implemented and supported in terms of their future versions and releases, and we expressed our opinion that such a concept will be a critical success factor for cloud-based software engineering to take off.

10   Conclusion

We have shown how to extend an object constraint language with reflection. Reflective constraint writing is to constraint writing what generative programming is to programming. We have extended the concrete object constraint language OCL of the UML modeling language stack for this purpose, resulting in so-called OCL_R. We have shown how to give precise, declarative semantics for OCL_R on the basis of semantical reification operators Φ and Ψ that mitigate between the M2-, M1- and M0-levels of the meta level architecture.

As a by-product, we have shown how to generalize OCL property call expressions by a truly generative version. This means, we have shown how to generalize OCL property call expressions of the form $o.p$ to multi-class, multi-property call expressions of the most general kind $\{o_1 : C_1, .., o_n : C_n\}.\{p_1, .., p_m\}$, i.e., so that the classes C_i can be dynamically generated and properties p_i may be identified merely by name, i.e., may not be inherited from a common supertype.

First, reflective constraint writing can be exploited in quality assurance for system design. Then, a major goal of introducing OCL_R was to support the analysis of semantics and pragmatics of modeling constructs. Another goal of reflective constraint writing is to enable sustainable constraints, which are, typically, constraints involving meta-level access. We have clarified why sustainable constraint writing is important for a robust modeling process. As an example, we have elaborated sustainable constraints, i.e., constraints that persist model evolution, for the modeling of sets of sets.

We have shown how usual class diagrams are sufficient to adequately model sets of sets of domain objects – given that constraints are provided that are appropriately made robust against M1-level updates. We have introduced the concepts of subset-global attribute and subclass attribute. We have introduced and analyzed the subtype externalization pattern. We have introduced and analyzed the power type externalization pattern. The two patterns of subtype externalization and power type externalization open a design space. We have discussed advantages and disadvantages of each of the modeling alternatives. The fact that even basic OO modeling languages are not reductionist as compared to, e.g., the PD (Parsimonious Data) modeling language in form-oriented analysis [26, 36], once more shows the conceptual redundancy of subtype externalization and power type externalization. We have defined and analyzed power type construction as a diamond consisting out of subtype externalization and power type externalization.

We have achieved precise semantics for conceptual models of arbitrarily nested sets. We have argued that the definition of power type in the UML specification is inconsistent. Based on the findings and terminology of this article, a precise re-formulation of the above definition has been possible. We have given a precise specification of the UML power types semantics with OCL_R.

We distinguished three viewpoints onto today's information systems, i.e., the classical viewpoint, the symbolic viewpoint and the purely symbolic viewpoint. It is the purely symbolic viewpoint that has served best to explain the potential of emerging multilevel modeling tools as evolving data storing systems.

Reflective constraint writing adds value. Reflective constraint writing can make constraints robust against model updates. There are many use cases for reflective constraints in different software engineering domains, i.e., both in system design and conceptual modeling. With respect to system design, reflective constraints can be exploited to ensure better artifact quality. They can be used, e.g., to enforce style guides or the correct application of design patterns. Conceptually, reflective constraint writing is about the externalization of important domain knowledge that is otherwise captured in the ephemeral counterparts of sustainable constraints.

Acknowledgements. I am grateful to Roland Wagner and Josef Küng for the many inspiring discussions on the foundations and, in particular, on the realization of database information systems, e.g., in the context of the DEXA series of conferences and related events. In particular, I am also grateful for the joint endeavors in distributed workflow automation projects.

A   UML Meta Model

The class diagram in Fig. 11 shows a cutout of the UML superstructure specification [75] consisting of all UML meta model elements used in the OCL constraints in this article. We have repeated some of classes, i.e., TypedElement and Feature, for the sake of improving overall readability.

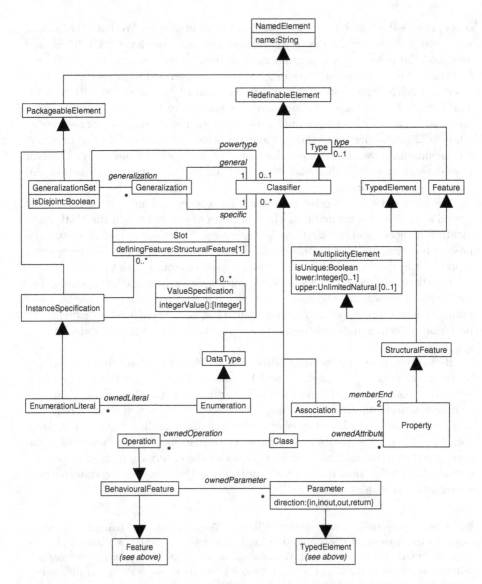

Fig. 11. Cutout of the UML meta model (superstructure) as needed in this article.

B OCL Types Abstract Syntax

Figure 12 shows the abstract syntax of the OCL v2.0 types. Basically, it shows the types from Figure 8.1 from the OCL v2.0 specification [70]. The singleton *AnyType* meta object *OclAny* serves as most general type for all OCL expressions, i.e., $e :: OCLExpression$ implies $e : AnyType$. The singleton *TypeType* meta object *OclType* serves as type for all OCL type expressions, i.e., $e :: TypeExpr$

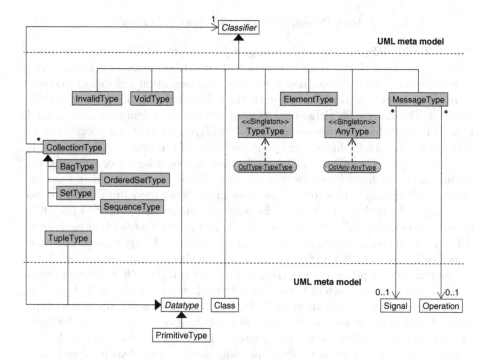

Fig. 12. Meta model of OCL v2.0

implies $e : OclType$. Model elements that are genuine to the OCL type specification are given in gray color, whereas, meta model elements that are reused from the UML superstructure specification [75] have white color.

B.1   On the Choice of OCL version v2.0

We have chosen to take the 2006 version OCL v2.0 [70] instead of the current version v2.4 [73] and the ISO standard version v2.3.1 [51,72] as the basis for OCL_R. The reason is the type system. The crucial difference is in the existence of the type $OclType$ and its corresponding abstract syntax element $Type Type$ which are present in the former version v2.0 but absent from the newer versions. The type $OclType$ is needed for a complete definition of well-typing. It serves as type for type expressions, i.e., for expressions $e :: TypeExp$ – see Appendix 13. For example, the problem shows in the definition of the property $oclIsTypeOf$. The property is defined in and OCL v2.0 and the newer OCL version in different ways:

$$oclIsTypeOf(type : OclType) : Boolean \qquad (83)$$

$$oclIsTypeOf(type : Classifier) : Boolean \qquad (84)$$

The operation applys to all objects, i.e., objects $o : OclAny$. It test whether the object's type equals the type given as parameter. For example, the following constraint evaluates to true:

$$\textbf{context } Person \textbf{ inv}: self.oclIsTypeOf(Person) \qquad (85)$$

The OCL v2.0 definition (83) of *oclIsTypeOf* is correct, whereas the v2.0 definition (84) cannot is ill-typed with respect to its described semantics. Even worse, in accordance with its described semantics, the operation *oclIsTypeOf(type : Classifier)* cannot be typed at all with the types avalailble in the newer OCL versions. The expression *Person* is a type expression that denotes a user-defined type. In v2.0 this expression has type *OclType*, so that the definition of *oclIsTypeOf* is correct. Let's turn to the definition of the newer OCL versions. It states that *type : Classifier*. The type *Classifier* can only be a user-defined type, among the pre-defined types there is no type *Classifier*. Here is where the misunderstanding might stem from. The types in the meta model in Fig. 12 are no OCL types themselves. They yield the abstract syntax that describes the OCL types. The existence of the class *Class* in the meta model means that each user-defined type serves as an OCL type, i.e., as a type for OCL expressions. The class *Class* itself is not an OCL type. And so is not the abstract class *Classifier*. Now, the semantic description requires the parameter of *oclIsTypeOf* is a type expression and not an expression of user-defined type. This means that *oclIsTypeOf* is ill-typed in the newer versions of OCL. Furthermore there is no appropriate type available in the newer version of OCL that could be given to the parameter *type*. In v2.0 the type *OclType* serves this purpose. The type *OclType* – yet without a defining abstract syntax and a corresponding meta model element *TypeType* – has been available in OCL since its first 1997 version OCL v1.1 and disappeared from the OCL specification in 2010 with version OCL v2.2 [71].

B.2   Flattening OCL Collections

In the OCL, nested collections are automatically flattened. Each combination of nested collections yields a concrete flattened collection which is defined in [70,70]. We have turned the definition of this flattening into a combinator $\oplus$ for collection constructors – see the definition in Table 2.

The standard fixes concrete results for the combination of collection into nested structures. Actually, there is a design space. Of course, it is natural to turn a set of sets into a set and, similarly, to turn a bag of bags into a bag. However, with respect to the combination of bags and sets the OCL has taken a deliberate decision for a symmetric solution, i.e., a bag of sets is turned into a bag, whereas a set of bags is turned into a set. This means, that in the latter case,

Table 2. The collection type combinator $\oplus$.

_ $\oplus$ _ (T)	Set	Bag	Sequence
Set	Set(T)	Set(T)	Set(T)
Bag	Bag(T)	Bag(T)	Bag(T)
Sequence	Sequence(T)	Sequence(T)	Sequence(T)

some information is lost, that is inherent in the encompassed bags. Furthermore, the construction of sequences out of sets and bags is not straightforward, in the OCL it is solved non-deterministically.

C OCL Expressions Abstract Syntax

Figure 13 gives a substantial cutout of the OCL abstract syntax as specified in [72]. Basically, it shows, as a single overview, the structure of the OCL syntax kernel as given in Fig. 8.2. in [72] plus more elements that are crucial for understanding the syntax and semantics of OCL_R, in particular, Fig. 13 details out the abstract syntax of feature call expressions. Model elements that are genuine to the OCL meta model are given in gray color, whereas, meta model elements that are reused from the UML superstructure specification [75] have white color.

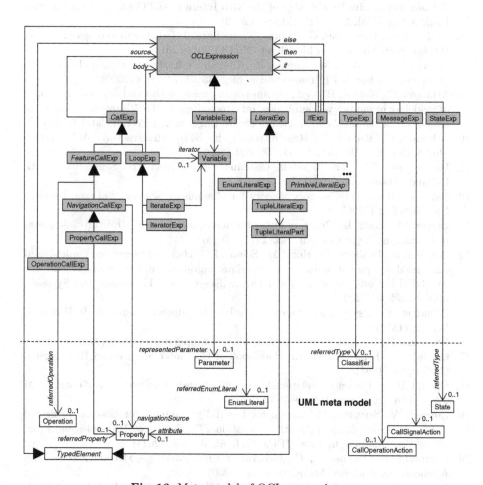

Fig. 13. Meta model of OCL expressions.

References

1. Abadi, M., Cardelli, L.: A Theory of Objects. Springer, New York (1996)
2. Asikainen, T., Männistö, T.: Nivel – a metamodeling language with a formal semantics. Softw. Syst. Model. **8**(4), 521–549 (2009)
3. Atkinson, C., Bostan, P., Draheim, D.: Foundational MDA patterns for service-oriented computing. J. Object Technol. **13**(5), 1–30 (2015)
4. Atkinson, C., Bostan, P., Draheim, D.: A unified conceptual framework for service-oriented computing. In: Hameurlain, A., Küng, J., Wagner, R. (eds.) TLDKS VII. LNCS, vol. 7720, pp. 128–169. Springer, Heidelberg (2012)
5. Atkinson, C., Draheim, D.: Cloud aided-software engineering - evolving viable software systems through a web of views. In: Mahmood, Z., Saeed, S. (eds.) Software Engineering Frameworks for the Cloud Computing Paradigm, pp. 255–281. Springer, London (2013)
6. Atkinson, C., Gerbig, R., Kennel, B., General-purpose, symbiotic, languages, domain-specific. In: Proceedings of the 34th International Conference on Software Engineering, ICSE 2012. IEEE Press (2012)
7. Atkinson, C., Grossman, G., Kühne, T., de Lara, J. (eds.): Proceedings of the 1st Workshop on Multi-Level Modelling, MULTI 2014 (2014)
8. Atkinson, C., Gutheil, M., Kennel, B.: A flexible infrastructure for multilevel language engineering. IEEE Trans. Softw. Eng. **35**(6), 742–755 (2009)
9. Atkinson, C., Kennel, B., Goß, B.: Supporting constructive and exploratory modes of modeling in multi-level ontologies. In: Proceedings of the 7th International Conference on Semantic Web-Enabled Software Engineering, SWESE 2011 (2011)
10. Atkinson, C., Kühne, T.: Rearchitecting the UML infrastructure. ACM Trans. Model. Comput. Simulat. **12**(4), 290–321 (2002)
11. Barendregt, H.P.: The Lambda Calculus - Its Syntax and Semantics. North Holland, Amsterdam (1984)
12. Barr, M., Wells, C.: Category Theory for Computing Science, 2nd edn. Prentice Hall, Reading (1995)
13. Berger, M., Tratt, L.: Program logics for homogeneous generative run-time meta-programming. Logic Comput. Sci. **11**(5) (2015)
14. Bordbar, B., Draheim, D., Horn, M., Schulz, I., Weber, G.: Integrated model-based software development, data access and data migration. In: Proceedings of the 8th ACM/IEEE Conference on Model Driven Engineering, Languages and Systems, MODELS 2005 (2005)
15. Brentano, F.: Psychologie vom empirischen Standpunkt. Duncker & Humblot, Leipzig (1874)
16. Brentano, F.: Psychology from an Empirical Standpoint. Routledge, London (1995)
17. Cardelli, L.: Type systems. In: Handbook of Computer Science and Engineering. CRC Press (1997)
18. Chen, P.P.-S.: The entity-relationship model - toward a unified view of data. ACM Trans. Database Syst. **1**(1), 9–36 (1976)
19. Corradi, A., Servetto, M., Zucca, E.: DeepFJig: modular composition of nested classes. In: Proceedings of the 9th International Conference on Principles and Practice of Programming in Java, PPPJ 2011. ACM Press (2011)
20. Czarnecki, K., Eisenecker, U.: Generative Programming - Methods, Tools, and Applications. Addison-Wesley, Reading (2000)
21. Deridder, D., et al. (eds.): Pre-proceedings of the International Workshop on Models and Evolution, MODELS 2011 (2010)

22. van Deursen, A., Klint, P., Visser, J.: Domain-specific languages: an annotated bibliography. ACM SIGPLAN Not. **35**(6), 26–36 (2000)
23. de Lara, J., Guerra, E.: Deep meta-modelling with METADEPTH. In: Vitek, J. (ed.) TOOLS 2010. LNCS, vol. 6141, pp. 1–20. Springer, Heidelberg (2010)
24. de Lara, J., Guerra, E.: Generic meta-modelling with concepts, templates and mixin layers. In: Petriu, D.C., Rouquette, N., Haugen, Ø. (eds.) MODELS 2010, Part I. LNCS, vol. 6394, pp. 16–30. Springer, Heidelberg (2010)
25. Draheim, D., Weber, G.: Strongly typed server pages. In: Halevy, A.Y., Gal, A. (eds.) NGITS 2002. LNCS, vol. 2382, pp. 29–44. Springer, Heidelberg (2002)
26. Draheim, D., Weber, G.: Modeling submit/response style systems with form charts and dialogue constraints. In: Meersman, R., Tari, Z. (eds.) OTM 2003. LNCS, vol. 2889, pp. 267–278. Springer, Heidelberg (2003)
27. Draheim, D.: Business Process Technology - A Unified View on Business Processes, Workflows and Enterprise Applications. Springer, Heidelberg (2010)
28. Draheim, D., Horn, M., Schulz, I.: The schema evolution, data migration framework of the environmental mass database IMIS. In: Proceedings of the 16th International Conference on Scientific and Statistical Database Management. IEEE (2004)
29. Draheim, D.: Sustainable constraint writing and a symbolic viewpoint of modeling languages. In: Decker, H., Lhotská, L., Link, S., Spies, M., Wagner, R.R. (eds.) DEXA 2014, Part I. LNCS, vol. 8644, pp. 12–19. Springer, Heidelberg (2014)
30. Draheim, D., Himsl, M., Jabornig, D., Leithner, W., Regner, P., Wiesinger, T.: Intuitive visualization-oriented metamodeling. In: Bhowmick, S.S., Küng, J., Wagner, R. (eds.) DEXA 2009. LNCS, vol. 5690, pp. 727–734. Springer, Heidelberg (2009)
31. Draheim, D., Himsl, M., Jabornig, D., Küng, J., Leithner, W., Regner, P., Wiesinger, T.: Concept and pragmatics of an intuitive visualization-oriented meta-modeling tool. J. Vis. Lang. Comput. **21**(4), 157–170 (2010). Elsevier
32. Draheim, D., Natschlger, C.: A context-oriented synchronization approach. In: Proceedings of the 2nd International Profile Management, and Context Awarness, VLDB Workshop in Personalized Access, PersDB 2008 (2008)
33. Draheim, D., Lutteroth, C., Weber, G.: Generative programming for C#. ACM SIGPLAN Not. **40**(8), 29–33 (2005). ACM Press
34. Draheim, D., Lutteroth, C., Weber, G.: A type system for reflective program generators. In: Glück, R., Lowry, M. (eds.) GPCE 2005. LNCS, vol. 3676, pp. 327–341. Springer, Heidelberg (2005)
35. Draheim, D., Lutteroth, C., Weber, G., Factory: statically type-safe integration of genericity and reflection. In: Proceedings of the 4th International Conference on Software Engineering, Artificial Intelligence, Networking, and Parallel/Distributed Computing, ACIS 2003 (2003)
36. Draheim, D., Weber, G.: Modelling form-based interfaces with bipartite state machines. J. Interact. Comput. **17**(2), 207–228 (2005). Elsevier
37. Draheim, D., Weber, G.: Form-Oriented Analysis - A New Methodology to Model Form-Based Applications. Springer, Heidelberg (2005)
38. Froehner, T., Nickles, M., Weiss, G.: Open ontologies – the need for modeling heterogeneous knowledge. In: Proceedings of the International Conference on Information and Knowledge Engineering, IKE 2004 (2004)
39. Gosling, J., Joy, B., Steele, G., Bracha, G.: The Java Language Specification, 3rd edn. Addison Wesley, Reading (2005)
40. Gurevich, Y.: Evolving Algebras 1993 - Lipari Guide. Oxford University Press, New York (1995)

41. Guarino, N.: Formal ontology and information systems. In: Guarino, N. (ed.) Proceedings of the 1st International Conference on Formal Ontology and Information Systems, pp. 3–15. IOS Press, Amsterdam (1998)
42. Hoare, C.A.R.: An axiomatic basis for computer programming. Commun. ACM **12**(10), 30–32 (1969)
43. Hayes, I.J.: Applying formal specification to software development in industry. IEEE Trans. Softw. Eng. **11**(2), 169–178 (1985)
44. Hayes, I.J.: Specification Case Studies. Prentice Hall, London (1993)
45. Henderson-Sellers, B.: Bridging metamodels and ontologies in software engineering. J. Syst. Softw. **84**, 303–313 (2011)
46. Henderson-Sellers, B., Gonzalez-Perez, C.: Connecting powertypes and stereotype. J. Object Technol. **4**(7), 83–96 (2005). ETH, Zürich
47. Henderson-Sellers, B., Gonzalez-Perez, C.: The rationale of powertype-based meta-modelling to underpin software development methodologies. In: Proceeding of the 2nd Asia-Pacific Conference on Conceptual Modelling, APCCM 2005, vol. 43, Australian Computer Society (2005)
48. Himsl, M., Jabornig, D., Leithner, W., Regner, P., Wiesinger, T., Küng, J., Draheim, D.: An iterative process for adaptive meta- and instance modeling. In: Wagner, R., Revell, N., Pernul, G. (eds.) DEXA 2007. LNCS, vol. 4653, pp. 519–528. Springer, Heidelberg (2007)
49. ISO: ISO/IEC 13568:2002. Information technology – Z Formal Specification Notation – Syntax, Type System and Semantics. ISO (2002)
50. ISO: International Standard ISO/IEC 24744: Software Engineering – Metamodel for Development Methodologies. ISO (2007)
51. ISO: Information Technology - Object Management Group Object Constraint Language, version 2.3.1, ISO Standard ISO/IEC 19507: (E), ISO (2012)
52. Johnson, R., Woolf, B.: Type object. In: Pattern Languages of Program Design, vol. 3. Addison-Wesley (1997)
53. Kappel, G., Preishuber, S., Prö II, E., Rausch-Schott, S., Retschitzegger, W., Wagner, R., Gierlinger, C.: COMan – coexistence of object-oriented and relational technology. In: Loucopoulos, P. (ed.) ER 1994. LNCS, vol. 881, pp. 259–277. Springer, Heidelberg (1994)
54. Kolyang, T., Santen, B., Wolff, B.: A structure preserving encoding of Z in Isabelle/HOL. In: Goos, G., Hartmanis, J., van Leeuwen, J., von Wright, J., Grundy, J., Harrison, J. (eds.) TPHOLs 1996. LNCS, vol. 1125, pp. 283–298. Springer, Heidelberg (2007)
55. Kühne, T., Schreiber, D.: Can programming be liberated from the two-level style? Multi-level programming with DeepJava. In: Proceedings of the 22th ACM SIGPLAN Conference on Object-Oriented Programming, Systems, Languages, and Applications, OOPSLA 2007. ACM Press (2007)
56. Kühne, T.: Matters of metamodeling. Softw. Syst. Model. **5**(1), 369–385 (2006). Springer
57. Lutteroth, C., Draheim, D., Weber, G.: A type system for reflective program generators. Sci. Comput. Program. **76**(5), 392–422 (2011). Elsevier
58. Martin-Löf, P.: Intuistionistic Type-Theory. Bibliopolis (1984)
59. Mahmood, Z., Saeed, S. (eds.): Software Engineering Frameworks for Cloud Computing Paradigm. Springer, London (2013)
60. Martin, J., Odell, J.J.: Object-Oriented Methods - A Foundation (UML). Prentice Hall, Englewood Cliffs (1998)
61. Morgan, C.: Programming from Specification. Prentice Hall, Englewood Cliffs (1990)

62. Neumayr, B., Grün, K., Schrefl, M.: Multi-level domain modeling with M-objects and M-relationships. In: Proceedings of the 6th Asia-Pacific Conference on Conceptual Modeling, APCCM 2009. Australian Computer Society (2009)
63. Neumayr, B., Schrefl, M.: Comparison criteria for ontological multi-level modeling. Presented at: Dagstuhl Seminar on The Evolution of Conceptual Modeling, Technical Report 08.03., Johannes-Kepler-University Linz (2008)
64. Nipkow, T., Paulson, L.C., Wenzel, M.: Isabelle/HOL - A Proof Assistant for Higher-Order Logic. Springer, Heidelberg (2002)
65. Nystrom, N., Qi, X., Myers, A.C.: J& – nested intersection for scalable software composition. In: Proceedings of the 21th ACM SIGPLAN Conference on Object-Oriented Programming, Systems, Languages, and Applications, OOPSLA 2006. ACM Press (2006)
66. Nystrom, N., Chong, S., Myers, A.C.: Scalable extensibility via nested inheritance. In: Proceedings of the 19th ACM SIGPLAN Conference on Object-Oriented Programming, Systems, Languages, and Applications, OOPSLA 2004. ACM Press (2004)
67. Odell, J.J.: Dynamic and multiple classification. J. Object Oriented Program. 4(9), 45–48 (1998)
68. Odell, J.J.: Power types. J. Object Oriented Program. 7(2), 8–12 (1994)
69. OMG: Object Constraint Language, version 1.1. Rational Software Corporation et al. (1997)
70. OMG: Object Constraint Language, version 2.0, OMG (2006)
71. OMG: Object Constraint Language, version 2.2, OMG (2010)
72. OMG: Object Constraint Language, version 2.3.1, OMG (2012)
73. OMG: Object Constraint Language, version 2.4, OMG (2014)
74. OMG: OMG Unified Modeling Language - Infrastructure, version 2.4.1. OMG (2011)
75. OMG: OMG Unified Modeling Language - Superstructure, version 2.4.1. OMG (2011)
76. OMG: OMG Meta Object Facility - Core Specification, version 2.4.1. OMG (2011)
77. Parr, T., Fisher, K.: LL(*) - the foundation of the ANTLR parser generator. In: Proceedings of the 32nd ACM SIGPLAN Conference on Programming Language Design and Implementation, PLDI 2011. ACM Press (2011)
78. Parnas, D.L.: A technique for software module specification with examples. Commun. ACM 15(5), 330–336 (1972)
79. Pierce, B.C.: Types and Programming Languages. MIT Press, Cambridge (2002)
80. Plotkin, G.: LCF considered a programming language. Theoret. Comput. Sci. 5, 229–256 (1977)
81. Razavi, R., Bouraqadi, N., Yoder, J.W., Perrot, J.-F., Johnson, R.E.: Language support for adaptive object-models using metaclasses. Comput. Lang. Syst. Struct. 31(3–4), 199–218 (2005)
82. Rossini, A., de Lara, J., Guerra, E., Rutle, A., Wolter, U.: A formalisation of deep metamodelling. Formal Aspects Comput. 26(6), 1115–1152 (2014)
83. Santen, T.: On the semantic relation of Z and HOL. In: Bowen, J.P., Fett, A., Hinchey, M.G. (eds.) ZUM 1998. LNCS, vol. 1493, pp. 96–116. Springer, Heidelberg (1998)
84. Schätz, B., et al. (eds.): Pre-proceedings of the International Workshop on Models and Evolution, MoDELS 2010 (2011)
85. Selic, B.: On the semantic foundations of standard UML 2.0. In: Bernardo, M., Corradini, F. (eds.) SFM-RT 2004. LNCS, vol. 3185, pp. 181–199. Springer, Heidelberg (2004)

86. Simons, P., Niemel, I., Soininen, T.: Extending and implementing the stable model semantics. Artif. Intell. **138**(1–2), 181–234 (2002)
87. Spivey, J.M.: The Z Notation. Prentice Hall, Englewood Cliffs (1992)
88. Lara, J., Vangheluwe, H.: Using AToM as a meta CASE tool. In: Proceedings of the 4th International Conference on Enterprise Information Systems, ICEIS 2002 (2002)
89. Yoder, J., Johnson, R.: The adaptive object model architectural style. In: Proceedings of the 3rd Working IEEE/IFIP Conference on Software Architecture, WICSA 2002. IEEE Press (2002)
90. Zermelo, E.: Untersuchungen über die Grundlagen der Mengenlehre. Mathematische Annalen **65**, 261–281 (1908)

PPP-Codes for Large-Scale Similarity Searching

David Novak[⊠] and Pavel Zezula

Masaryk University, Brno, Czech Republic
{david.novak,zezula}@fi.muni.cz

Abstract. Many current applications need to organize data with respect to mutual similarity between data objects. A typical general strategy to retrieve objects similar to a given sample is to access and then refine a *candidate set* of objects. We propose an indexing and search technique that can significantly reduce the candidate set size by combination of several space partitionings. Specifically, we propose a mapping of objects from a generic metric space onto main memory codes using several *pivot spaces*; our search algorithm first ranks objects within each pivot space and then *aggregates* these rankings producing a candidate set reduced by *two orders of magnitude* while keeping the same answer quality. Our approach is designed to well exploit contemporary HW: (1) larger main memories allow us to use rich and fast index, (2) multi-core CPUs well suit our parallel search algorithm, and (3) SSD disks without mechanical seeks enable efficient selective retrieval of candidate objects. The gain of the significant candidate set reduction is paid by the overhead of the candidate ranking algorithm and thus our approach is more advantageous for datasets with expensive candidate set refinement, i.e. large data objects or expensive similarity function. On real-life datasets, the search time speedup achieved by our approach is by factor of two to five.

1 Introduction

The complexity and diversity of digital data is permanently increasing, which naturally generates new requirements for data retrieval. For many contemporary data types, it is convenient or even essential that the access methods be based on mutual similarity of the data objects because it corresponds to the human perception of the data or because exact matching would be too restrictive (various multimedia, biomedical or sensor data, etc.). We adopt a generic approach to this problem, where the data space is modeled by a data domain $\mathcal{D}$ and a general *metric* function δ to assess dissimilarity between pairs of objects from $\mathcal{D}$.

The field of metric-based similarity search has been studied for almost two decades [29]. The general objective of metric accesses methods (MAMs) is to preprocess the indexed dataset $\mathcal{X} \subseteq \mathcal{D}$ in such a way that, given a query object $q \in \mathcal{D}$, the MAM can effectively identify objects x from $\mathcal{X}$ with the shortest distances $\delta(q, x)$. A good motivating example for our work is an image search based on visual similarity of the image content. Recent advances in the area of deep neural networks allow to "extract" a semantically rich visual feature from a digital image [9,17]; these features are 4096-dimensional float vectors (16 KB

© Springer-Verlag Berlin Heidelberg 2016
A. Hameurlain et al. (Eds.): TLDKS XXIV, LNCS 9510, pp. 61–87, 2016.
DOI: 10.1007/978-3-662-49214-7_2

each vector). The search domain $\mathcal{D}$ is then the feature space and the similarity function δ is Euclidean distance or other vector-based distance. For instance, if the data collection $\mathcal{X}$ contains 10 million objects, the goal of the MAM is to organize 160 GB of the feature data and answer queries like "find me a dozen images from the collection that are the most similar to this query image".

Current MAMs designed for large data collections are typically approximate [25,29] and adopt the following high-level approach: Dataset $\mathcal{X}$ is split into *Partitions*; given a query, partitions with the highest "likeliness" to contain query-relevant data are read from the disk and this data form the *candidate set* of objects x to be *refined* by explicit evaluation of $\delta(q, x)$. The search costs of this schema consist mainly of (1) the I/O costs of reading the candidate partitions from the disk (they can be accessed as continuous data chunks) and (2) CPU costs of refinement; thus, the overall costs of the search typically strongly correlate with the candidate set size.

In this work, we propose a technique that can significantly reduce the candidate set size. In complex data spaces, the data partitions often span relatively large areas of the space and thus the candidate set is either large or imprecise. The key idea of our approach is to use several independent space partitionings; given a query, each of these partitionings generates a ranked set of candidate objects and we propose a way to aggregate these rankings so that the resulting candidate set is small and precise. Only objects identified in this way are actually retrieved from the disk and refined. Specifically, our approach works as follows:

- The data space is partitioned using a set of *pivots* (reference objects, anchors) where position of each data object is determined by its *closest pivots* and their order; this defines a mapping of the data into a *pivot space*. We use such pivot spaces to partition the dataset independently multiple times. In this way, each object is mapped onto a code denoted as *PPP-Code*;
  its size can be adjusted so that codes of the whole dataset fit into the main memory.
- Given a query, we first rank the object codes within each pivot spaces with respect to the query. Further and more importantly, we propose a way to aggregate these several rankings, which provably increases the probability that the query-relevant objects appear high in the final ranking.
- The PPP-Codes are organized by an indexing structure, which can lower their memory occupation. This index is also used by the proposed PPPRANK algorithm that efficiently calculates individual candidate rankings and their aggregation; the algorithm exploits principles introduced by Fagin et al. [13].

This approach was designed with the idea of making the best use of contemporary trends in hardware development: (1) Larger main memories allow to maintain a rich memory index, (2) multi-core CPU architectures well support our demanding but accurate candidate set identification, and (3) SSD disks without mechanical seeks allow efficient retrieval of the candidate objects from the disk one-by-one (not as continuous data chunks).

Combination of independent partitionings was proposed before by LSH approaches [15] and it was also recognized by a few MAMs as a way to increase answer quality [11,23]; these works propose to simply replicate the data in multiple indexes and to *widen* the query candidate set by union of multiple independent candidate sets. On the contrary, our aggregation mechanism *shrinks* the candidate set significantly while maintaining the same answer quality.

The experiments conducted on three diverse datasets show that this approach can reduce the candidate set size by *two orders of magnitude*. The response times depend on the time spared by this candidate set reduction (reduced I/O costs and δ-refinement time) versus the overhead of the PPPRANK algorithm. To analyze this tradeoff, we have run experiments on an artificial dataset with adjustable object sizes and tunable time of δ evaluation; the results show that our approach is not worthwhile only for the smallest data objects with the fastest δ function. Most of the evaluations were realized on two real-life datasets (100M CoPhIR [6] and 1M complex visual signatures [5]); for these, our approach was two- to five-times faster than competitors on the same HW platform.

The paper is further organized as follows. In Sect. 2, we define fundamental terms and analyze current approaches; in Sect. 3, we propose the PPP-Encoding (Sect. 3.1), ranking within individual pivot spaces (Sect. 3.2) and rank aggregation (Sects. 3.3 and 3.4); Sect. 4 describes our index and search algorithm. Our approach is evaluated and compared with others in Sect. 5 and the paper is concluded in Sect. 6 with a reference to an online demonstration application built with the aid of the proposed technique. This work is an extension of a paper presented at DEXA 2014 [24].

2   Preliminaries and Related Work

We focus on indexing and searching based on mutual object distances and we primarily assume that the data is modeled as a metric space [29]:

Definition 1. Metric space *is an ordered pair* $(\mathcal{D}, \delta)$, *where* $\mathcal{D}$ *is a domain of objects and* δ *is a total* distance function $\delta : \mathcal{D} \times \mathcal{D} \longrightarrow \mathbb{R}$ *satisfying postulates of non-negativity, identity, symmetry, and triangle inequality.*

Our technique does not explicitly demand triangle inequality. In general, the metric-based techniques manage the dataset $\mathcal{X} \subseteq \mathcal{D}$ and search it by the *nearest neighbors query* K-NN(q), which returns K objects from $\mathcal{X}$ with the smallest distances to given $Q \in \mathcal{D}$ (ties broken arbitrarily). We assume that the search answer A may be an approximation of the precise K-NN answer A^P and the result quality is measured by $recall(A) = precision(A) = \frac{|A \cap A^P|}{K} \cdot 100\,\%$.

During two decades of research, many approximate metric access methods (MAMs) have been proposed [25,29]. Further in this section, we focus especially on (1) techniques based on the concept of *pivot permutations*, (2) approaches that use several independent space partitionings, and (3) techniques that propose memory encoding of data objects. Having a set of k pivots $P = \{p_1, \ldots, p_k\} \subseteq \mathcal{D}$, Π_x is a *pivot permutation* defined with respect to object $x \in \mathcal{D}$ iff $\Pi_x(i)$ is the index of the i-th closest pivot to x; accordingly, sequence $p_{\Pi_x(1)}, \ldots, p_{\Pi_x(k)}$ is ordered with respect to distances between the pivots and x (ties broken by order of the increasing pivot index). Formally:

Definition 2. *Having a set of k pivots $P = \{p_1, \ldots, p_k\} \subseteq \mathcal{D}$ (reference objects) and an object $x \in \mathcal{D}$, let Π_x be permutation on $\{1, \ldots, k\}$ such that $\forall i : 1 \leq i < k$:*

$$\delta(x, p_{\Pi_x(i)}) < \delta(x, p_{\Pi_x(i+1)}) \vee$$
$$(\delta(x, p_{\Pi_x(i)}) = \delta(x, p_{\Pi_x(i+1)}) \wedge \Pi_x(i) < \Pi_x(i+1)).$$

Π_x *will be referred to as* pivot permutation *(PP) with respect to x.*

Several techniques based on this principle [7,10,11,20] use the PPs to group data objects together (data partitioning); given a query, relevant partitions are read from the disk and refined; the relevancy is assessed based on the PPs. Unlike these methods, the MI-File [1] builds inverted file index according to object PPs; these inverted files are used to rank the data according to a query and the candidate set is then refined by accessing the objects one-by-one [1]. In this respect, our approach adopts similar principle and we compare our results with the MI-File (see Sect. 5.3).

In this work, we propose to use *several* independent pivot spaces (sets of pivots) to define several PPs for each data object and to identify candidate objects. The idea of multiple indexes is known from the Locality-sensitive Hashing (LSH) [15] and it was also applied by a few metric-based approaches [11,23]; some metric indexes actually define families of *metric LSH functions* [22]. All these works benefit from enlarging the candidate set by a simple union of the top results from individual indexes; on the contrary, we propose such rank aggregation that can significantly reduce the size of the candidate set in comparison with a single index while preserving the same answer quality. Recently, the C2LSH technique [14] proposed a way to combine LSH functions resulting in a partially ranked candidate set; this work aims mainly at vector spaces with known families of LSH functions and it does not assume any pre-ranking of candidate sets from individual indexes.

Several recent works focused on reducing the size of data by *source coding* (or quantization) of Euclidean vector spaces so that the codes fit into memory; the authors use approaches like unsupervised machine learning [27], spectral hashing [28], or product quantization [16] to define data objects codes together with new ranking methods on these codes. A purely distance-based approach was also proposed [18]; it uses k pairs of pivots, that divide the space by k generalized hyperplanes, and each i-th bit of the k-bit code of $x \in \mathcal{D}$ reflects on which side of the i-th hyperplane object x lies. From a high perspective, this approach is similar to ours, but we propose techniques for indexing and ranking and corresponding non-exhaustive search algorithm.

3   PPP-Encoding and Ranking

In this section, we introduce the principal ideas of our approach: (1) the PPP-Encoding of the data (Sect. 3.1), (2) ranking within individual pivot spaces (Sect. 3.2), and (3) aggregation of these rankings which defines the overall ranking of the PPP-Codes (Sect. 3.3). Section 3.4 contains basic effectiveness evaluation of the proposed ranking.

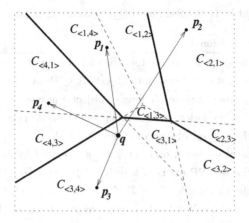

Fig. 1. Recursive Voronoi partitioning ($k = 4$, $l = 2$) and query-pivot distances.

3.1 Encoding by Pivot Permutation Prefixes

For a data domain $\mathcal{D}$ with distance function δ, object $x \in \mathcal{D}$ and a set of k pivots $P = \{p_1, \ldots, p_k\}$, the pivot permutation (PP) Π_x is defined as in Definition 2. In our technique, we do not use the full PP but only its prefix, i.e. the ordered list of a given number of nearest pivots:

Notation: Having pivots $\{p_1, \ldots, p_k\}$ and PP Π_x, $x \in \mathcal{D}$, we denote $\Pi_x(1..l)$ the pivot permutation prefix (PPP) of length l: $1 \le l \le k$, specifically

$$\Pi_x(1..l) = \langle \Pi_x(1), \Pi_x(2), \ldots, \Pi_x(l) \rangle. \tag{1}$$

The pivot permutation prefixes have a geometrical interpretation important for the similarity search – the PPPs actually define *recursive Voronoi partitioning* of the metric space [26]. Let us explain this principle on an example in Euclidean plane with four pivots $p_1, \ldots, p_4$ in Fig. 1; the thick solid lines depict borders between standard *Voronoi cells* – sets of points $x \in \mathcal{D}$ for which specific pivot p_i is the closest one: $\Pi_x(1) = i$. The dashed lines further partition these cells using other pivots; these sub-areas are labeled $C_{\langle i,j \rangle}$ and they cover all objects for which $\Pi_x(1) = i$ and $\Pi_x(2) = j$, thus $\Pi_x(1..2) = \langle i, j \rangle$.

Notation: For an l-tuple $\langle i_1, \ldots, i_l \rangle$, we denote $C_{\langle i_1, \ldots, i_l \rangle}$ the *Voronoi cell of level* l that contains all objects $x \in \mathcal{D}$ for which $\Pi_x(1..l) = \langle i_1, \ldots, i_l \rangle$.

The pivot permutation prefixes (PPPs) $\Pi_x(1..l)$ form the base of the proposed PPP-Encoding, which is composed of several PPPs for each object. Thus, let us further assume having λ independent sets of k pivots $P^1, P^2, \ldots, P^\lambda$, $P^j = \{p_1^j, \ldots, p_k^j\}$. For any $x \in \mathcal{D}$, each of these sets generates a PP Π_x^j, $j \in \{1, \ldots, \lambda\}$ and we can define the PPP-Encoding as follows.

Table 1. Notation used throughout this paper.

Symbol	Definition
$(\mathcal{D}, \delta)$	the data domain and metric distance $\delta : \mathcal{D} \times \mathcal{D} \to \mathbb{R}$
$\mathcal{X}$	the set of indexed data objects $\mathcal{X} \subseteq \mathcal{D}$; $|\mathcal{X}| = n$
k	number of pivots (reference objects) in one pivot space
$\Lambda = \{1, \ldots, \lambda\}$	Λ is the index set of λ independent pivot spaces
$P^j = \{p_1^j, \ldots, p_k^j\}$	the j-th set of k pivots from $\mathcal{D}$; $j \in \Lambda$
Π_x^j	pivot permutation of $(1 \ldots k)$ ordering P^j by distance from $x \in \mathcal{D}$
$\Pi_x^j(1..l)$	the j-th PP prefix of length l: $\Pi_x^j(1..l) = \langle \Pi_x^j(1), \ldots, \Pi_x^j(l) \rangle$
$PPP_l^{1..\lambda}(x)$	the PPP-Code of $x \in \mathcal{D}$: $PPP_l^{1..\lambda}(x) = \langle \Pi_x^1(1..l), \ldots, \Pi_x^\lambda(1..l) \rangle$
$C_{\langle i_1, \ldots, i_l \rangle}^j$	Voronoi cell of level l containing $x \in \mathcal{X}$ s.t. $\Pi_x^j(1..l) = \langle i_1, \ldots, i_l \rangle$
d, d_K, d_Δ	measures ranking pivot permutation prefixes (PPPs) $d(q, \Pi(1..l))$
$\psi_q^j : \mathcal{X} \to \mathbb{N}$	the j-th ranking of objects according to $Q \in \mathcal{D}$ generated by d
$\Psi_{\mathbf{p}}(q, x)$	the overall rank of x by the **p**-percentile of its $\psi_q^j(x)$ ranks, $j \in \Lambda$
R	size of candidate set – number of objects x refined by $\delta(q, x)$

Definition 3: *Having λ sets of k pivots and parameter $l : 1 \leq l \leq k$, we define* PPP-Code *of object $x \in \mathcal{D}$ as a λ-tuple*

$$PPP_l^{1..\lambda}(x) = \langle \Pi_x^1(1..l), \ldots, \Pi_x^\lambda(1..l) \rangle. \tag{2}$$

Individual components (PPPs) of the PPP-Code will be also denoted as $PPP_l^j(x) = \Pi_x^j(1..l)$, $j \in \{1, \ldots, \lambda\}$; to shorten the notation, we set $\Lambda = \{1, \ldots, \lambda\}$. These and other symbols used throughout this paper are summarized in Table 1.

The PPP-Encoding is exemplified in Fig. 2 where each of the $\lambda = 2$ pivot sets defines an independent Voronoi partitioning of the data space. Every object $x \in \mathcal{X}$ is encoded by $PPP_l^j(x) = \Pi_x^j(1..l)$, $j \in \Lambda$. Object x_5 is depicted in both diagrams and, for instance, within the first partitioning, the closest pivots from x_5 are $p_7^1, p_4^1, p_8^1, p_5^1$, which corresponds to $PPP_4^1(x_5) = \Pi_{x_5}^1(1..4) = \langle 7, 4, 8, 5 \rangle$.

3.2 Ranking of Pivot Permutation Prefixes

Having objects from $\mathcal{X}$ encoded as described above, we want to find ranking mechanism of $PPP_l^{1..\lambda}(x)$, $x \in \mathcal{X}$ with respect to query $q \in \mathcal{D}$, which would be an approximation of the ranking generated by distances $\delta(q, x)$. To achieve this, we first define rankings on components of $PPP_l^{1..\lambda}(x)$, prefixes $\Pi_x(1..l)$. In the following, we define and compare two such ranking measures d_K and d_Δ.

Measures Based Purely on Permutations. A natural approach is to project the query object q to the same space as the data objects by encoding q into its PP Π_q (or prefix $\Pi_q(1..l)$) and to calculate "distance" between Π_q and $\Pi_x(1..l)$.

first pivot space partitioning ($j = 1$)   second pivot space partitioning ($j = 2$)

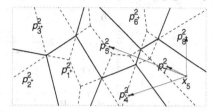

Fig. 2. Principles of encoding data objects as PPP-Codes $PPP_l^{1..\lambda}(x)$ with two pivot sets ($\lambda = 2$) each with eight pivots ($k = 8$) and using pivot permutation prefixes of length four ($l = 4$). Each object x is encoded by $PPP_4^{1..2}(x) = \langle \Pi_x^1(1..4), \Pi_x^2(1..4) \rangle$; the figure shows example of object x_5: $PPP_4^{1..2}(x_5) = \langle \langle 7, 4, 8, 5 \rangle, \langle 7, 8, 4, 6 \rangle \rangle$.

There are several standard ways to measure difference between full permutations that were also used in similarity search: Spearman Footrule, Spearman Rho or Kendall Tau measure [2,7]; the last mentioned seems to slightly outperform the others [7]. The Kendall Tau between permutations Π_x and Π_y defines for every pair $\{i, j\}$, $0 \leq i, j \leq k$: $K_{i,j}(\Pi_x, \Pi_y) = 0$ if indexes i, j are in Π_x *in the same order* as in Π_y; otherwise, we set $K_{i,j}(\Pi_x, \Pi_y) = 1$. The Kendall Tau is then defined as [12]:

$$K(\Pi_x, \Pi_y) = \sum_{1 \leq i, j \leq k} K_{i,j}(\Pi_x, \Pi_y).$$

This measure can be generalized in several ways to work with permutation *prefixes*, where not all $K_{i,j}$ are known [12]. We propose a measure d_K which calculates a distance between the query object $q \in \mathcal{D}$ and $\Pi(1..l)$ as *minimum* of Kendall Tau distances between the full permutation Π_q and all permutations Π' on $\{1, \ldots, k\}$ that have $\Pi(1..l)$ as prefix [12]:

$$d_K(q, \Pi(1..l)) = \min_{\Pi': \Pi'(1..l) = \Pi(1..l)} K(\Pi_q, \Pi'). \tag{3}$$

The accuracy of this measure with respect to the original distance δ is evaluated later in this section. There exists an algorithm for computation of the full Kendall Tau with $O(k \cdot \log k)$ complexity [8]. The same idea can be used to design an $O(l \cdot \log l)$ algorithm for d_K on permutation prefixes.

Measures that Use Query-Pivot Distances. The query object $q \in \mathcal{D}$ in the ranking function can be represented more richly than by permutation Π_q, specifically, we can use directly the query-pivot distances $\delta(q, p_1), \ldots, \delta(q, p_k)$; see Fig. 1, which depicts such distances.

If we consider only the first-level Voronoi cells ($l = 1$), thus only the closest pivots $\Pi_x(1)$, we can approximate the distance between a query and objects in cell $C_{\Pi(1)}$ by distance $\delta(q, p_{\Pi(1)})$ (this idea was described in [16]); for instance in Fig. 1, distance between q and cell $C_{\langle 4 \rangle}$ (delimited by the thick solid lines around

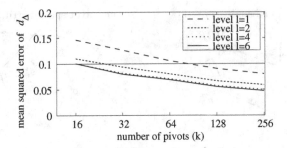

Fig. 3. Mean squared error of d_Δ (4) with $c = 0.75$ on 1M CoPhIR dataset.

pivot p_4) would be $\delta(q, p_4)$. Having the cells further partitioned according to other pivots, we propose to shift the distance estimation towards the next pivots that define cell $C_{\Pi(1..l)}$; influence of these next pivots should be smaller than of the first one. For instance, distances between q and cells $C_{\langle 4,1 \rangle}$, $C_{\langle 4,3 \rangle}$ would be (weighted) averages between $d(q, p_4)$ and $d(q, p_1)$, $d(q, p_3)$, respectively, which should make the estimations more precise.

Formally, we propose to measure the distance between q and $\Pi(1..l)$ as a *weighted arithmetic mean* of distances between q and the l pivots from $\Pi(1..l)$:

$$d_\Delta(q, \Pi(1..l)) = \left(\sum_{i=1}^{l} c^{i-1} \delta(q, p_{\Pi(i)}) \right) / \sum_{i=1}^{l} c^{i-1}, \qquad (4)$$

where c is parameter $0 < c \le 1$ to control the influence of the next pivots; for now, we set $c = 0.75$ and its influence is properly evaluated in Sect. 5.1. Naturally, this heuristic does not improve the distance estimation in *all* cases, but we consider the average influence. We measure the precision of the distance estimator as *mean squared error* [16] defined as

$$\text{MSDE}(d_\Delta) = \iint (\delta(q, x) - d_\Delta(q, \Pi_x(1..l)))^2 p(x)\mathrm{d}x\, p(q)\mathrm{d}q$$

where $p(\cdot)$ is the probability distribution function of the data domain $\mathcal{D}$, $q, x \in \mathcal{D}$. Figure 3 depicts the values of $\text{MSDE}(d_\Delta)$ measured by Monte-Carlo sampling (averages over a large set of samples) on the CoPhIR dataset (see Sect. 5 for description of the dataset). The graph shows development of MSDE as the space partitioning is refined by growing number of pivots k and by increasing PPP length l used by d_Δ. We can see that levels $l > 1$ can improve MSDE so that k would have to be multiplied to achieve such MSDE values for $l = 1$. The same trends were observed for all other datasets (Sect. 5).

Comparison of the PPP Ranking Measures. Let us briefly compare effectiveness of the measures d_K (3) and d_Δ (4) proposed in the previous two subsections. In the following, d will stand for any dissimilarity measure between $q \in \mathcal{D}$ and $\Pi(1..l)$ such as d_K or d_Δ. Such measures d together with $q \in \mathcal{D}$ naturally

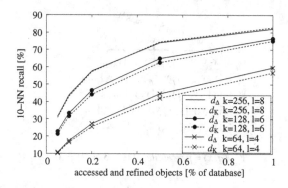

Fig. 4. Recall of 10-NN on 1M CoPhIR dataset as accessing up to 1% of the data ordered according to d_K and d_Δ; $k = 64, 128, 256, l = 4, 6, 8$.

induce ranking ψ_q of the indexed set $\mathcal{X}$ according to growing distance from q. Formally, ranking $\psi_q : \mathcal{X} \to \mathbb{N}$ is the smallest numbering of set $\mathcal{X}$ that fulfills the following condition for all $X, y \in \mathcal{X}$:

$$d(q, \Pi_x(1..l)) \leq d(q, \Pi_y(1..l)) \Rightarrow \psi_q(x) \leq \psi_q(y). \tag{5}$$

We define the effectiveness of measure d as average recall of K-NN if the dataset $\mathcal{X}$ is accessed in the order ψ_q generated by d. We have compared this effectiveness on several datasets and with various settings; Fig. 4 shows graphs of the 10-NN recall on a 1M subset of the CoPhIR dataset (see Sect. 5) as up to 1% of this set is accessed according to d_K and d_Δ with several selected parameters of k and l.

These graphs well illustrate the generally observed trend: d_Δ is slightly better for smaller values of k and l (and also for very small numbers of accessed objects) while for extremely fine-grained space partitioning, both measures have practically the same effectiveness. These results are in compliance with previous works [20]. Regarding the lower complexity of d_Δ ($\Theta(l)$) in comparison to d_K ($O(l \cdot \log l)$), we choose d_Δ as the measure used in the rest of this paper.

3.3 Aggregation of Multiple Rankings

At this point, we know how to rank objects $x \in \mathcal{X}$ with respect to $q \in \mathcal{D}$ within one pivot space. As above, we assume that this ranking $\psi_q(x)$ (5) is induced by measure d such as d_K (3) or d_Δ (4) applied on q and PPPs $\Pi_x(1..l)$. Let us now assume that x is encoded by $PPP_l^{1..\lambda}(x)$ codes composed of λ such PPPs and that we have a mechanism able to provide λ sorted lists of objects $x \in \mathcal{X}$ generated by measure d between q and $\Pi_x^j(1..l)$, $j \in \Lambda$. Then, $\psi_q^j(x)$ denotes the position of x in the j-th ranking, $j \in \Lambda$. Figure 5 (top part) shows an example of five rankings ψ_q^j, $j \in \{1, \ldots, 5\}$.

These rankings are partial – objects with the same PPP $\Pi(1..l)$ have the same rank (objects from the same recursive Voronoi cell). This is the main

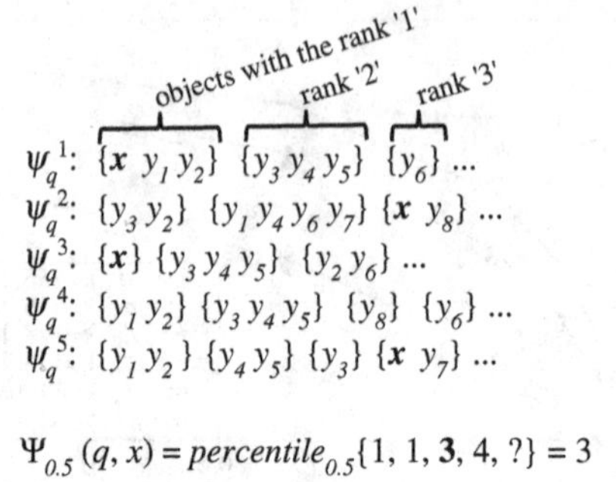

$$\overbrace{}^{\text{objects with the rank '1'}} \overbrace{}^{\text{rank '2'}} \overbrace{}^{\text{rank '3'}}$$

$\psi_q^1:\ \{x\ y_1\ y_2\}\ \{y_3\ y_4\ y_5\}\ \{y_6\}\ \dots$

$\psi_q^2:\ \{y_3\ y_2\}\ \{y_1\ y_4\ y_6\ y_7\}\ \{x\ y_8\}\ \dots$

$\psi_q^3:\ \{x\}\ \{y_3\ y_4\ y_5\}\ \{y_2\ y_6\}\ \dots$

$\psi_q^4:\ \{y_1\ y_2\}\ \{y_3\ y_4\ y_5\}\ \{y_8\}\ \{y_6\}\ \dots$

$\psi_q^5:\ \{y_1\ y_2\}\ \{y_4\ y_5\}\ \{y_3\}\ \{x\ y_7\}\ \dots$

$$\Psi_{0.5}(q, x) = percentile_{0.5}\{1, 1, 3, 4, ?\} = 3$$

Fig. 5. Rank aggregation by $\Psi_\mathbf{p}$ of object $x \in \mathcal{X}$, $\lambda = 5$, $\mathbf{p} = 0.5$.

source of inaccuracy of these rankings because, in complex data spaces, the Voronoi cells typically span relatively large areas and thus the top positions of ψ_q contain both objects close to q and more distant ones. Having several independent partitionings, the query-relevant objects should be at top positions of most of the rankings while the "noise objects" should vary because the Voronoi cells are of different shapes. The objective of our rank aggregation is to filter out these noise objects. Namely, we propose to assign each object $x \in \mathcal{X}$ the $\mathbf{p}$-percentile of its ranks, $0 \le \mathbf{p} \le 1$:

$$\Psi_\mathbf{p}(q, x) = percentile_\mathbf{p}(\psi_q^1(x), \psi_q^2(x), \dots, \psi_q^\lambda(x)). \tag{6}$$

For instance, $\Psi_{0.5}$ assigns median of the ranks; see Fig. 5 for an example – positions of object x in individual rankings are: 1, 3, 1, *unknown*, 4 and median of these ranks is $\Psi_{0.5}(q, x) = 3$. This principle was used by Fagin et al. [13] for a different purpose and they propose MEDRANK algorithm for efficient calculation of $\Psi_\mathbf{p}$. This algorithm does not require to explicitly find out *all* ranks of a specific object, but only $\lceil \mathbf{p}\lambda \rceil$ first (best) ranks (this is explicit in Fig. 5). Details and properties of the MEDRANK algorithm [13] are provided in Sect. 4.

Now, we would like to show that the $\Psi_\mathbf{p}$ aggregation actually improves the ranking in comparison with a single ψ_q ranking by increasing the probability that objects close to q will be assigned top positions (and vice versa). Also, we would like to find theoretically suitable values of $\mathbf{p}$.

Let x be an object from the dataset $\mathcal{X}$ and p_z be the probability such that $p_z = Pr[\psi_q(x) \le z]$, where $z \ge 1$ is a position in ψ_q ranking. Having λ independent rankings $\psi_q^j(x)$, $j \in \Lambda$, we want to determine probability $Pr[\Psi_\mathbf{p}(q, x) \le z]$ with respect to p_z. Let X be a random variable representing the number of ψ_q^j ranks of x that are smaller than z: $|\{\psi_q^j(x) \le z, j \in \Lambda\}|$. Assuming that the probability distribution of p_z is the same for each of $\psi_q^j(x)$, we get

$$Pr[X = j] = \binom{\lambda}{j} \cdot (p_z)^j \cdot (1 - p_z)^{\lambda - j}.$$

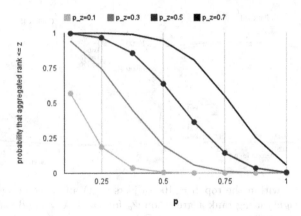

Fig. 6. Development of $Pr[\Psi_\mathbf{p}(q,x) \leq z]$ for $\lambda = 8$, selected p_z and variable $\mathbf{p}$.

In order to have $\Psi_\mathbf{p}(q,x) \leq z$, at least $\lceil \mathbf{p}\lambda \rceil$ positions of x must be $\leq z$ and thus

$$Pr[\Psi_\mathbf{p}(q,x) \leq z] = \sum_{j=\lceil \mathbf{p}\lambda \rceil}^{\lambda} Pr[X = j].$$

for the variable $\mathbf{p}$ and selected values of p_z ($\lambda = 8$). We can see that the aggregation increases the differences between individual levels of p_z (for non-extreme $\mathbf{p}$ values); e.g. for $\mathbf{p} = 0.5$, probabilities $p_z = 0.1$ and $p_z = 0.3$ are transformed to lower probability values whereas $p_z = 0.5$ and $p_z = 0.7$ are pushed to higher probabilities. The probability $p_z = Pr[\psi_q(x) \leq z]$ naturally grows with z but, more importantly, we assume that p_z is higher for objects close to q then for distant ones. Because ψ_q^j are generated by distance between q and Voronoi cells (5) and these cells may be large, there may be many distant objects that appear at top positions of individual ψ_q although having low probability p_z. The rank aggregation $\Psi_\mathbf{p}(q,x)$ for non-extreme $\mathbf{p}$ values can push away such objects and increase the probability that top ranks are assigned only to objects close to q (Fig. 6).

Table 2. Sequential scan experiment parameters.

Parameter	Description	Default
λ	number of pivot spaces	4
k	pivot number in each space	128
l	length of PPP	8
$\mathbf{p}$	percentile used in $\Psi_\mathbf{p}$	0.75
R	candidate set size	100 (0.01 %)

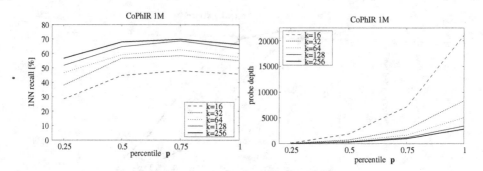

Fig. 7. 1NN recall within the top $R = 100$ objects (left) and average probe depth of each ψ_q^j, $j \in \Lambda$ (right) using rank aggregation $\Psi_{\mathbf{p}}$ for $l = 8$, $\lambda = 4$ and various number of pivots k and percentile $\mathbf{p}$.

3.4 Accuracy of the PPP-Encoding and Ranking

Let us evaluate the basic accuracy of the K-NN search if the objects are encoded by PPP-Codes and ranked by $\Psi_{\mathbf{p}}(q, x)$. Results in this section are independent of any indexing and searching algorithms, namely, we use the sequential scan and focus entirely on the trends and mutual influence of several parameters summarized in Table 2; all results are on the 1M CoPhIR dataset (see Sect. 5). We measure the accuracy as K-NN recall within the top R candidate objects $x \in \mathcal{X}$ identified by $\Psi_{\mathbf{p}}(q, x)$. In this section, we present results of 1-NN recall, which has the same trend as other values of K. All results are averaged over 1,000 randomly selected queries outside the dataset and all pivot sets P^j were selected independently at random from the dataset.

Graphs in Fig. 7 focus on the influence of percentile $\mathbf{p}$. The left graph shows average 1-NN recall within the top $R = 100$ objects for variable $\mathbf{p}$ and selected k. We can see that, as expected, the higher k the better and, more importantly, the peak of the results is at $\mathbf{p} = 0.75$ (just for clarification, for $\lambda = 4$, $\Psi_{0.75}(q, x)$ is equal to the third $\psi_q^j(x)$ rank of x out of four). These measurements are in compliance with the expectations discussed in the previous section.

The right graph in Fig. 7 shows the *probe depth* [13] – the average number of objects that had to be accessed in each ranking $\psi_q^j(x)$, $j \in \Lambda$ in order to discover $R = 100$ objects in at least $\lceil \mathbf{p}\lambda \rceil$ rankings (and thus determine their $\Psi_{\mathbf{p}}(q, x)$). Naturally, the probe depth grows with $\mathbf{p}$, especially for $\mathbf{p} \geq 0.75$. We can also see that finer space partitioning (higher k) results in a lower probe depth because the Voronoi cells are smaller and thus objects close to q appear in $\lceil \mathbf{p}\lambda \rceil$ rankings sooner. The general lessons learned are: the more pivots the better (for both recall and efficiency), ideal percentile seems to be around 0.5–0.75, which is in compliance with results of Fagin et al. [13].

In general, we can assume that accuracy of the ranking will grow with increasing values of k, l, and λ, but these parameters influence the size of the PPP-code representation of an object:

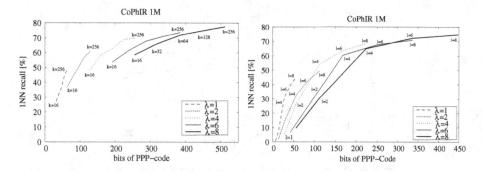

Fig. 8. 1NN recall within the top $R = 100$ objects ranked by $\Psi_{0.75}$ for selected λ and various bit length of $PPP_l^{1..\lambda}(x)$ influenced either by number of pivots k (left, $l = 8$) or by prefix length l (right, $k = 128$).

$$\text{size of } PPP_l^{1..\lambda}(x) = \lambda \cdot l \cdot \lceil \log_2 k \rceil \text{ bits.} \tag{7}$$

Figure 8 shows recall for the variable bit size of $PPP_l^{1..\lambda}$ codes for selected values of λ. In the left graph, the prefix length is fixed at $l = 8$ and the code size is influenced by increasing number of pivots k; we can see that higher values of recall can be achieved only with larger λ but, on the other hand, considering the PPP-Code lengths, it is more convenient to increase k than λ. The right graph presents a similar experiment, only here $k = 128$ and parameter l is increased; we can see that it pays off to increase l than to increase λ. The recall improvement achieved by increasing k and l is practically the same (with respect to PPP-Code size); higher k means more query-pivot distance computations.

The graph in Fig. 9 adopts an inverse point of view – it answers the question how the aggregation approach reduces the candidate set size R necessary to achieve given recall (80 %, in this case); please, notice the logarithmic scales. The small numbers in the graph show the reduction factor with respect to $\lambda = 1$; we can see that R is reduced down to about 5 % using $\lambda = 8$.

4 Indexing of PPP-Codes and Search Algorithm

So far, we have proposed a way to encode metric objects by PPP-Codes and to rank these codes according to given query object. In this section, we propose (1) an index to be built on a PPP-encoded dataset that can *decrease the memory footprint* of the PPP-Codes, and (2) an efficient non-exhaustive search algorithm.

4.1 Dynamic PPP-Tree Index

The $PPP_l^{1..\lambda}(x)$ code is composed of λ PPPs $\Pi_x(1..l)$. Given a set of these l-tuples, some of them would share common prefixes of variable lengths. In order to spare memory, we propose a *PPP-Tree* index – a dynamic trie structure that would keep the l'-prefixes of the PPP-Codes only once for all objects sharing

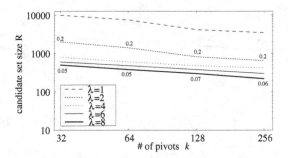

Fig. 9. Cand. set size R necessary for 80 % 1-NN recall.

the same l'-prefix, $l' \leq l$. This PPP-Tree index is to be built for each of the λ pivot spaces. Similar indexes were used in PP-Index [11] and in M-Index [20], but without the objective of memory representation reduction.

Schema of the PPP-Tree index is sketched in Fig. 10. Intuitively, let $i_1, \ldots, i_{l'}$ be indexes on a path from the root of the tree to a certain node at level l'. This node and its subtree contains all objects $x \in \mathcal{X}$ for which $\Pi_x(1..l') = \langle i_1, \ldots, i_{l'} \rangle$. This node also corresponds to Voronoi cell $C_{\langle i_1, \ldots, i_{l'} \rangle}$ (see Sect. 3.1). An *internal* node at level $l' < l$ is composed of these entries:

$$\langle i_{l'+1}, ptr \rangle, \quad \text{where } i_{l'+1} \in \{1, \ldots, k\} \setminus \{i_1, \ldots, i_{l'}\};$$

pointer ptr points at subtree containing objects with PPP $\langle i_1, \ldots, i_{l'}, i_{l'+1} \rangle$; possible values of index $i_{l'+1}$ are limited because the indexes must be unique within a permutation. Let us recall that $\Pi_x(l'+1..l)$ denotes the part of the pivot permutation of object x between positions $l'+1$ and l; further, let $\oplus$ denote concatenation of two tuples. A *leaf* node at level l' is composed of entries

$$\langle \Pi_x(l'+1..l), \mathrm{ID}_x \rangle,$$

where ID_x is the unique identifier of object $x \in \mathcal{X}$ for which $\Pi_x(1..l) = \langle i_1, \ldots, i_{l'} \rangle \oplus \Pi_x(l'+1..l)$. Entries in leaves at level $l' = l$ degenerate to $\langle \langle \rangle, \mathrm{ID}_x \rangle$ where $\Pi_x(1..l) = \langle i_1, \ldots, i_l \rangle$.

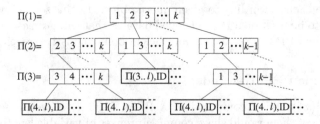

Fig. 10. Schema of a single dynamic PPP-Tree.

The key to memory efficiency of such a structure is its dynamic leveling. Splitting a leaf node with n' objects at level l', $1 \leq l' < l$ spares $n' \cdot \lceil \log_2 k \rceil$ bits because the memory representation of $\Pi_x(l'+1..l)$ would be shorter by one index if these n' objects are moved to level $l' + 1$. On the other hand, the split creates new leaves with certain overhead; thus we propose to split the leaf iff

$$n' \cdot \lceil \log_2 k \rceil > b \cdot \text{NODEOVERHEAD}$$

where b is potential branching of the leaf, $b \leq n'$ and $b \leq k - l' + 1$. The actual value of b can be either precisely measured for each leaf or estimated based on the statistics of average branching at level l'. Value of NODEOVERHEAD depends on implementation details.

So far, we have described a single PPP-Tree (as if $\lambda = 1$). Having $\lambda > 1$, we propose to build a separate PPP-Tree for each $j \in \Lambda$. In this case, an object x in all λ trees is "connected" by its ID_x stored in the leaf cells of the trees. This generates additional memory overhead per data object in comparison with sequential scan, because $\lambda - 1$ additional IDs need to be stored. We consider that an identifier ID has $\lceil \log_2 n \rceil$ bits for dataset $|\mathcal{X}| = n$.

4.2   Non-Exhaustive Search Algorithm

In Sect. 3.3, we have proposed a way to aggregate λ rankings of indexed objects $\mathcal{X}$ and we have briefly mentioned the MEDRANK algorithm [13]. Having the PPP-Tree indexes as described above, we can now propose the PPPRANK algorithm that does this aggregation effectively. The main procedure (Algorithm 1) follows the idea of MEDRANK; the PPP-Tree structures are used for effective generation of individual λ rankings (subroutine GETNEXTIDS).

Given a query object $q \in \mathcal{D}$, percentile $0 \leq \mathbf{p} \leq 1$ and number R, PPPRANK returns IDs of R indexed objects $x \in \mathcal{X}$ with the lowest value of $\Psi_{\mathbf{p}}(q, x)$; please, recall that this aggregated rank is defined as the $\lceil \mathbf{p}\lambda \rceil$-th best position from $\psi_q^j(x)$ ranks, $j \in \Lambda$ (6). In every iteration (lines 4–9), the algorithm accesses next objects of all rankings (routine GETNEXTIDS(q, j), $j \in \Lambda$); set S carries the already seen objects x together with the number of their occurrences in the rankings (frequencies f_x). GETNEXTIDS(q, j) always returns next object(s) with the best ψ_q^j rank and thus, when an object x achieves frequency $f_x \geq \lceil \mathbf{p}\lambda \rceil$, it is guaranteed that any object y achieving $f_y \geq \lceil \mathbf{p}\lambda \rceil$ in a subsequent iteration of PPPRANK must have higher rank $\Psi_{\mathbf{p}}(q, y) > \Psi_{\mathbf{p}}(q, x)$ [13].

Idea of the GETNEXTIDS(q, j) subroutine is to traverse the j-th PPP-Tree using a priority queue Q. As we know, each PPP-Tree node corresponds to Voronoi cell $C_{\langle i_1, \dots, i_{l'} \rangle}$, $l' \leq l$; the queue Q is always ordered by $d(q, \langle i_1, \dots, i_{l'} \rangle)$ (where d is the measure that generates ranking ψ_q^j (5)). In every iteration, the head of Q is processed; if head is a leaf node, its objects identifiers ID_x are inserted into Q ranked by $d(q, \Pi_x^j(1..l))$. When object identifiers appear at the head of Q, they are returned as "next objects in the j-th ranking". Algorithm 2 formalizes this routine: Q is composed of triples $\langle dist, \langle i_1, \dots, i_{l'} \rangle, \_ \rangle$ where the third component is either a node N or an object ID. Initially, Q contains the

Algorithm 1. PPPRANK($q, \mathbf{p}, R$)

Input: $q \in \mathcal{D}$; percentile $\mathbf{p}$; candidate set size R
Output: IDs of R objects $x \in \mathcal{X}$ with lowest $\Psi_{\mathbf{p}}(q, x)$
// S is a set of ''seen objects'': ID_x with their frequencies f_x
1 set $S \leftarrow \emptyset$ // A is answer list of object IDs
2 list $A \leftarrow \langle \rangle$
3 **while** $|A| < R$ **do**
4   **foreach** $j \in \Lambda$ **do**
5   **foreach** ID_x in GETNEXTIDS(q, j) **do**
6   **if** $\mathrm{ID}_x \notin S$ **then**
7   add ID_x to S and
8   set $f_x = 1$
9   **else**
10   increment f_x
11   **foreach** ID_x in S $such$ $that$ $f_x \geq \lceil \mathbf{p}\lambda \rceil$ **do**
12   move ID_x from S to A
13 **return** A

---

root of the j-th PPP-Tree (line 2) and, in every step, the tree node at the head of Q (line 4) is decomposed and either its successors are inserted into Q (if N is internal, line 8) or the object IDs are put into Q (if N is a leaf, line 12). In both cases, the PP prefix of the successor node (or object) is reconstructed from the PP prefix of node N and information from the node entry. If object IDs appear at the head of Q, the top IDs with the same distance $d(q, \Pi_x^j(1..l))$ are returned (lines 13–18); these are IDs of objects with the same j-th rank (see Fig. 5).

Symbol d stands for a measure between $q \in \mathcal{D}$ and $\Pi(1..l)$ such as d_K (3) or d_Δ (4). The following property is key to correctness of Algorithm 2: For any PP Π and l', $1 \leq l' < l$:

$$d(q, \Pi(1..l')) \leq d(q, \Pi(1..l'+1)). \tag{8}$$

This property is fulfilled by d_K (3) but not by d_Δ (4); thus, we slightly modify Eq. (4) to calculate *weighted sum* of the query-pivot distances (instead of weighted average):

$$d_\Delta(q, \Pi(1..l)) = \sum_{i=1}^{l} c^{i-1} \delta(q, p_{\Pi(i)}). \tag{4'}$$

See Appendix I for correctness of Algorithm 2.

Complexity of GETNEXTIDS. This routine strongly influences efficiency of the whole PPPRANK algorithm. The amortized complexity of the nested loops in Algorithm 2 depends on the number of items inserted to the queue Q; the queue can be implemented as a binary heap and thus the whole complexity would be $O(|Q| \cdot \log |Q|)$. In an ideal case, Q would contain *only* the data IDs to be returned and the tree nodes on the path from the root to these data IDs. As Q is ordered by the d-distance to individual nodes, this case would require

---

Algorithm 2. GETNEXTIDS(q, j)

Input: query $q \in \mathcal{D}$, index $j \in \Lambda$
Output: IDs of the next objects in j-th ranking
1 calculate $\delta(q, p_i^j)$, $\forall i \in \{1, \ldots, k\}$
   // Q of triples $\langle dist, \langle i_1, .., i_{l'} \rangle, \_ \rangle$ ordered by $dist$
2 priority queue $Q \leftarrow \{\langle 0.0, \langle \rangle, \text{root of } j\text{-th PPP-Tree}\rangle\}$
   // end of INITIALIZATION (do once for each q, j)
3 **while** Q.head.3rd_component *is a tree node* **do**
4 $\quad \langle dist, \langle i_1, \ldots, i_{l'} \rangle, N \rangle \leftarrow Q$.dequeue()
5 $\quad$ **if** N *is an internal node* **then**
6 $\quad\quad$ **foreach** *entry* $\langle i_{l'+1}, ptr \rangle$ *in node* N **do**
7 $\quad\quad\quad \Pi_{ptr} \leftarrow \langle i_1, \ldots, i_{l'}, i_{l'+1} \rangle$
8 $\quad\quad\quad Q$.enqueue($\langle d(q, \Pi_{ptr}), \Pi_{ptr}, deref(ptr)\rangle$)
9 $\quad$ **else**
10 $\quad\quad$ **foreach** *entry* $\langle \Pi_x^j(l'+1..l), \mathrm{ID}_x \rangle$ *in node* N **do**
11 $\quad\quad\quad \Pi_x^j(1..l) = \langle i_1, \ldots, i_{l'} \rangle \oplus \Pi_x^j(l'+1..l)$
12 $\quad\quad\quad Q$.enqueue($\langle d(q, \Pi_x^j(1..l)), \Pi_x^j(1..l), \mathrm{ID}_x \rangle$)
   // return IDs of the top objects x in the queue
13 $A \leftarrow \emptyset$
14 **repeat**
15 $\quad \langle d_x, \Pi_x^j(1..l), \mathrm{ID}_x \rangle \leftarrow Q$.dequeue()
16 $\quad A \leftarrow A \cup \{\mathrm{ID}_x\}$
17 **until** Q.head.3rd_comp *is node* $\vee$ Q.head.1st_comp $> d_x$
18 **return** A

---

that *all other* tree nodes have distances larger than the d-distances to the returned IDs. Consequently, the length of Q depends on "tightness" of Eq. (8) – difference between the d-distance of an internal cell and the d-distance of its successors. See Appendix II for three algorithm optimizations that help to shorten the Q.

Search Process Review. Schema in Fig. 11 reviews the whole search process. Given a K-NN(q) query, the first step is calculating distances between q and all pivots: $\delta(q, p_i^j)$, $i \in \{1, \ldots, k\}$, $j \in \Lambda$. This is a necessary initialization of the GETNEXTIDS(q, j) procedures (steps 3), which generate the continual rankings ψ_q^j that are consumed by the main PPPRANK$(q, \mathbf{p}, R)$ algorithm (step 2). The candidate set of R objects x is retrieved from the disk (step 4) and refined by calculating $\delta(q, x)$ (step 5). The whole process can be parallelized in the following way: The λ steps 3 run fully in parallel and step 2 continuously reads their results; in this way, the full ranking $\Psi_\mathbf{p}(q, x)$ is generated item-by-item and is immediately consumed by steps 4 and then 5.

5 Efficiency Evaluation

We evaluate efficiency of our approach on three datasets; two of them are real-life, and the third one is artificially created to have fully controlled test conditions:

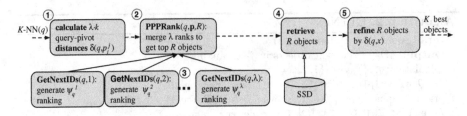

Fig. 11. Search pipeline using the PPP-Encoding and PPPRANK algorithm.

CoPhIR 100 million objects each consisting of five MPEG-7 global visual descriptors extracted from an image [6]. The distance function δ is a weighted sum of partial descriptor distances [3]; each object consumes 590 B on the disk (59 GB for 100M objects) and the computation of δ takes around 0.01 ms;

SQFD 1 million visual feature *signatures* each consisting of, on average, 60 cluster centroids in a 7-dimensional space; each cluster has a weight and such signatures are compared by *Signature Quadratic Form Distance* (SQFD) [5] which is a cheaper alternative to Earth Movers Distance. Each object occupies around 1.8 kB on disk and the SQFD distance takes around 0.5 ms;

ADJUSTABLE 10 million float vectors uniformly generated from $[0,1]^{32}$ compared by Euclidean distance; the disk size of each object can be artificially set from 512 B to 4096 B (5 GB to 40 GB for 10M objects) and time of δ computation can be tuned between 0.001 ms and 1 ms.

As a result of the analysis reported in Sect. 3.4, the indexes use these parameters: $l = 8$, $\lambda = 5$, $\mathbf{p} = 0.5$ (3 out of 5); the CoPhIR index uses $k = 256$, 384 and 512, SQFD index has $k = 64$, and the ADJUSTABLE index $k = 128$. The pivot sets P^j were selected independently at random from the datasets. As in Sect. 3.4, we use d_Δ (4') to generate individual ψ_q^j. The presented results are an average over 1,000 random K-NN(q) queries. The efficiency is gauged by standard measures from similarity search field [23, 29]:

I/O costs number of 4 kB block reads; in our approach, it is practically equal to the candidate set size R (step 4);

distance computations (DC) number of evaluations of distance δ; equal to $\lambda \cdot k + R$ (steps 1 and 5);

search time the wall-clock time of the search process running parallel as described above.

All experiments were conducted on a machine with 8-core Intel Xeon @ 2.0 GHz, 12 GB of RAM and a SATA SSD disk (CrystalDiskMark benchmark speed: sequential read 440 MB/s, random 4K QD32 read 270 MB/s); for comparison, we also present some results on the following HDD configuration: two 10,000 rpm magnetic disks in RAID 1 array (CrystalDiskMark sequential read 150 MB/s). All techniques used the full memory for their index and for disk caching; caches were cleared before every batch of 1,000 queries. The implementation is in Java using the MESSIF framework [4].

Table 3. Size of PPP-Code representation without index (sequential scan) and with the dynamic PPP-Tree Index.

Dataset	k	Single object + ID (no index)	Sequential scan occupation	Single object + IDs (with index)	Memory index occupation
SQFD	64	240 + 20 b	32 MB	161 + 100 b	32.5 MB
ADJUSTABLE	128	280 + 24 b	365 MB	217 + 120 b	403 MB
CoPhIR	256	320 + 27 b	4.2 GB	205 + 135 b	4.0 GB
	384	360 + 27 b	4.6 GB	245 + 135 b	4.5 GB
	512	360 + 27 b	4.6 GB	258 + 135 b	4.6 GB

5.1   PPP-Tree and PPPRank Overhead

Our approach encodes each object by a PPP-Code and a PPP-Tree index is built on these codes. Table 3 shows the sizes of this representation for individual datasets. The third column shows the size of the PPP-Code representation (7) plus the object ID size (unique within given dataset); the fourth column is the overall size of the sequential scan built on these PPP-Codes for given dataset. The next column shows PPP-Code sizes as reduced by PPP-Tree – in this case, the object IDs are stored λ-times (see Sect. 4.1); the last column shows the overall sizes of the PPP-Tree indexes. We can see that the memory reduction by PPP-Trees and increase by multiple ID storage are practically equal.

From now on, let us focus on the search efficiency. As mentioned above, the I/O costs and the number of δ computations (DC) are generated in steps 1, 4 and 5 of the search and can be directly derived from the algorithm parameters k, λ and R. Let us have a closer look at the costs of the PPPRANK algorithm itself – steps 2 and 3. Complexity of the aggregation step 2 depends directly on the *probe depth*, which was already mentioned in Sect. 3.4.

Figure 12 shows statistics of the PPPRANK algorithm on the full 100M CoPhIR dataset. The left graph shows the development of the probe depth (left axis) and the 10-NN recall (right axis) with respect to the weight c from the weighted sum d_Δ (4′). As we know, this weight influences individual $\psi_q^j(x)$, $j \in \Lambda$ and thus the result quality – we can see that the optimal recall is achieved around $c = 0.8$. It is interesting that the recall is in a perfect inverse correlation with the probe depth – if the PPPRANK needs to read fewer objects from individual $\psi_q^j(x)$ rankings, then the output objects are closer to q. This confirms the idea behind our aggregation approach.

The right graph in Fig. 12 shows length of the queue Q (left axis) that determines complexity of GETNEXTIDS. As analyzed in Sect. 4.2, the Q length depends on tightness of Eq. (8), which is directly influenced by parameter c; the response time (right axis) depends on the probe depth and on Q length. Considering these results, we further fixate $c = 0.75$ as a compromise between answer quality and efficiency.

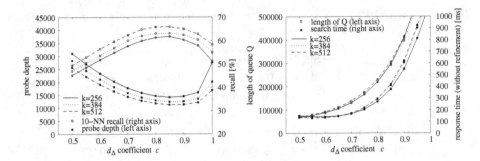

Fig. 12. PPPRANK on 100M CoPhIR varying coefficient c from d_Δ (4'); the candidate set size $R = 1000$. Left: probe depth and 10-NN recall; right: length of priority queue Q, response time (without steps 4 and 5).

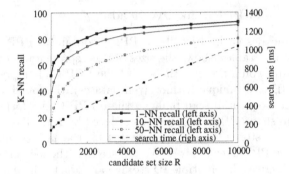

Fig. 13. Recall and search time on 100M CoPhIR as the candidate set grows; $k = 512$.

5.2 The Overall Efficiency

Naturally, the quality of the search result comes at the expense of higher search costs. In Sect. 3.4, we have studied influence of the code size (fineness of the partitioning adjusted at the building phase) to the answer recall, but the main parameter to increase the recall at query time is R (the size of the candidate set). Figure 13 shows development of recall (left axis) and search time (right axis) with respect to R on the CoPhIR dataset ($k = 512$). We can see that our approach can achieve very high recall while accessing thousands of objects out of 100M. The recall grows very steeply in the beginning, achieving about 90 % for 1-NN and 10-NN around $R = 5000$; the time grows practically linearly.

Table 4 presents more measurements on the CoPhIR dataset. We have selected two values of $R = 1000$ and $R = 5000$ (10-NN recall 64 % and 84 %, respectively) and we present the I/O costs, computational costs, and the overall search times on both SSD and HDD disks. All these results should be put in context – comparison with other approaches. At this point, let us mention metric structure M-Index [20], which is based on similar fundamentals as our approach: it computes a PPP for each object, maintains an index structure

Table 4. Results on 100M CoPhIR ($k = 512$) and SQFD ($k = 64$) with PPP-Codes and M-Index (512 and 64 pivots).

	technique	cand. set R	recall 10-NN	recall 50-NN	I/O costs	# of δ comp.	time on SSD [ms]	time on HDD [ms]
CoPhIR	PPP-Codes	1,000	65 %	47 %	1,000	3,560	270	770
		5,000	84 %	72 %	5,000	7,560	690	3,000
	M-Index	110,000	65 %	54 %	15,000	110,000	320	1,300
		400,000	85 %	80 %	59,000	400,000	1050	5,400
SQFD	PPP-Codes	100	70 %	45 %	100	420	160	230
		1,000	95 %	89 %	1,000	1,320	250	550
	M-Index	3,200	65 %	53 %	1,500	2,200	650	820
		13,000	94 %	89 %	6,100	6,500	750	920

Table 5. Search times [ms] of PPP-Codes / M-Index on ADJUSTABLE with 10-NN recall 85 %. PPP-Codes: $k = 128$, $R = 1000$, M-Index: 128 pivots, $R = 400000$.

PPP-Code / M-Index [ms]	size of an object [bytes]			
	512	1024	2048	4096
δ time 0.001 ms	370 / **240**	**370** / 410	**370** / 1,270	**370** / 1,700
0.01 ms	**380** / 660	**380** / 750	**380** / 1,350	**380** / 1,850
0.1 ms	**400** / 5,400	**400** / 5,400	**420** / 5,500	**420** / 5,700
1 ms	**1,100** / 52,500	**1,100** / 52,500	**1,100** / 52,500	**1,100** / 52,500

similar to our single PPP-Tree (Fig. 10), and it accesses the leaf nodes based on a scoring function similar to d_Δ; our M-Index implementation shares the core with the PPP-Codes and it stores the data in continuous disk chunks for efficient reading. Comparison of M-Index and PPP-Codes shows precisely the gain and the overhead of PPPRANK algorithm, which aggregates λ partitionings.

Looking at Table 4, we can see that M-Index with 512 pivots needs to access and refine $R = 110000$ or $R = 400000$ objects to achieve 65 % or 85 % 10-NN recall, respectively; the I/O costs and number of distance computations correspond with R. According to these measures, the PPP-Codes are one or two orders of magnitude more efficient than M-Index; looking at the search times, the dominance is not that significant because of the PPPRANK algorithm overhead. Please, note that the M-Index search algorithm is also parallel – both reading of the data and refinement are parallelized [20].

In order to clearly uncover the conditions under which the PPP-Codes overhead is worth the gain of reduced I/O and DC costs, we have introduced the ADJUSTABLE dataset. Table 5 shows the search times of PPP-Codes/M-Index while the object disk size and the DC time are adjusted. The results are measured on 10-NN recall level of 85 %, which is achieved at $R = 1000$ and $R = 400000$ for the PPP-Codes and M-Index, respectively; please, note that these values of R mean even more dramatic candidate set reduction observed for this uniformly distributed dataset than in case of CoPhIR. Looking at the search times at Table 5, we can see that for the smallest objects and fastest distance, the M-Index beats PPP-Codes but as values of these two variables grow, the PPP-Codes show their strength. We believe that this table well summarizes the overall strength and costs of our approach.

The SQFD dataset is an example of data type belonging to the lower middle part of Table 5 – the signature objects occupy almost 2 kB and the SQFD distance function takes 0.5 ms on average. A graph in Fig. 14 presents the PPP-Codes K-NN recall and search times while increasing R (note that size of the dataset is 1M and $k = 64$). We can see that the index achieves excellent results between $R = 500$ and $R = 1000$ with search time under 300 ms. Again, let us compare these results with M-Index with 64 pivots – the lower part of Table 4 shows that the PPP-Code aggregation can decrease the candidate set size R down under 1/10 of the M-Index results (for comparable recall values). For this dataset, we let the M-Index store precomputed object-pivot distances together with the objects and use them at query time for distance computation filtering [20, 29]; this significantly decreases its DC costs and search times, nevertheless, the times of PPP-Codes are about 1/5 of the M-Index.

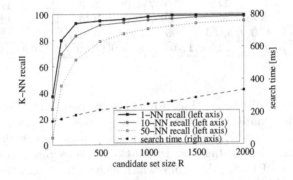

Fig. 14. Recall and search time on 1M SQFD dataset as candidate set grows; $k = 64$.

These results can be summarized as follows: The proposed approach is worthwhile for data types with larger objects (over 512 B) or with the time-consuming δ function (over 0.001 ms). For the two real-life datasets, our aggregation schema cuts the I/O and δ computation costs down by one or two orders of magnitude. The overall speed-up factor is about 1.5 for CoPhIR and 5 for the SQFD dataset.

5.3   Comparison with Other Approaches

Finally, let us compare our approach with selected relevant techniques for approximate metric-based similarity search. We focus especially on those works that present results on the full 100M CoPhIR dataset; the results on this dataset are summarized in Table 6 and analyzed below.

Table 6. Comparison with other approaches on 100M CoPhIR dataset.

Technique	Overall # of pivots	Cand. set R	Recall 10-NN	I/O costs	# of δ comp.
PPP-Codes	2,560	5,000	84 %	5,000	7,560
M-Index	512	400,000	85 %	59,000	400,512
M-Index (4 indexes)	960	300,000	84 %	44,000	301,000
PP-Index (8 indexes)	800	~333,000	86 %	~49,000	~334,000
	8,000	~52,000	82 %	~7670	~60,000
MI-File	20,000	1,000	88 %	~20,000	21,000

M-Index. We have described the M-Index [20] and compared it with our approach in the previous section, because it shares the core idea with PPP-Codes
which makes these approaches well comparable. We have chosen the M-Index
also because its scoring function seems to be at least as good [20] as of other
PP-based approaches [7,11]. The first two lines in Table 6 summarize the results
of PPP-Codes and M-Index on 100M CoPhIR. The third line shows variant when
four M-Indexes are combined by a standard union of the candidate sets [23]; we
can see that this approach can reduce the candidate set size R but the PPP-
Codes still outperform it significantly.

PP-Index. The PP-Index [11] also uses prefixes of pivot permutations to partition
the data space; it builds a slightly different tree structure on the PPPs, identifies
query-relevant partitions using a different heuristic, and reads these candidate
objects in a disk-efficient way. In order to achieve high recall values, the PP-
Index also combines several independent indexes by merging their results [11];
Table 6 shows selected results – we can see that the values are slightly better
than those of M-Index, especially when a high number of pivots is used (8,000)
but the PPP-Codes access less than 1/10 of the PP-Index candidate set.

MI-File. The MI-File [1] creates inverted files according to pivot-permutations;
at query time, it determines the candidate set, reads it from the disk one-by-one
and refines it. Table 6 shows selected results on 100M CoPhIR; we can see that
extremely high number of pivots (20,000) resulted in even smaller candidate set
then in case of PPP-Codes. The I/O costs are higher due to the disk size of the
MI-File index and the computational costs are higher due to a high number of
query-pivot δ distance evaluations.

As mentioned above, structures like PP-Index [11] or M-Index [20,23] use
multiple independent partitionings and they union candidate sets (or answers)
from them; this is also well known from the LSH approach [15]. Let us compare
the rank aggregation proposed in this work with the simple union of ranked
candidate sets from multiple partitionings. Figure 15 shows candidate set size R
necessary to achieve 80 % 1-NN recall when λ ranks generated by d_Δ are merged
either by $\Psi_{0.75}$ or by union (results are on 1M CoPhIR with the same settings
as in Sect. 3.4). We can see that the Ψ aggregation results in less than half R.

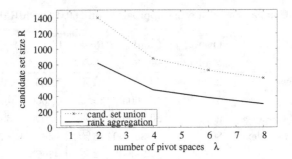

Fig. 15. Comparison of R using candidate set union vs. PPP-Codes rank aggregation; 1M CoPhIR, $k = 128$, $l = 8$, $\mathbf{p} = 0.75$, 80% 1-NN recall.

6   Conclusions

Efficient generic similarity search on a very large scale would have applications in many areas dealing with various complex data types. This task is difficult especially because identification of query-relevant data in complex data spaces typically requires accessing and refining relatively large candidate sets. If the data objects are large or if the refining similarity function is time-consuming then the search process may be unacceptably demanding. Since contemporary data types are often large and use complex similarity functions, we have designed a technique that would pay much more attention to identifying an accurate candidate set at the expense of higher algorithm complexity.

We have proposed a rich index by encoding each object using multiple pivot spaces; this PPP-Code index can be adjusted to fit into the main memory. Further, we have proposed a two-tier search algorithm – the first part of the algorithm generates several independent data object rankings according to distance between a query and the data codes, and the second part aggregates these rankings into one that provably increases the probability that query-relevant objects are accessed sooner.

We have conducted experiments on three datasets and in all cases our aggregation approach reduced the candidate set size by one or two orders of magnitude while preserving the answer quality. Because our search algorithm is relatively demanding, the overall search time gain depends on specific dataset. First, an artificial dataset with adjustable properties has helped us to show that our approach is not profitable only for data types with small objects and cheap similarity function. The second dataset was the 100M content-based image retrieval collection CoPhIR [6]; our approach speeded up the search on this set twice. Finally, we have used a dataset of 1M signature descriptors with a demanding SQFD distance function [5] where the candidate set reduction speeded up the search process more than five times.

Our approach differs from others in three aspects. First, it transfers a large part of the search computational burden from the similarity function evaluation towards the search process itself and thus the search times are very stable across different data types. Second, our index explicitly exploits a larger chunk of main memory in comparison with an implicit use for disk caching. And third, our approach reduces the I/O costs and it fully exploits the strength of the SSD disks without mechanical seeks or, possibly, of a fast distributed key-value store [19].

The PPP-Codes index forms the heart of an application that demonstrates a large-scale visual image search [21]. A collection of 20 million images has been processed by a deep convolutional neural network to obtain powerful visual features [9]. Compared by Euclidean distance, these 4096-dimensional vector features well express semantic similarity of digital images; uncompressed, 20M features occupy over 320 GB on the disk. The PPP-Codes index can reach a very good answer quality accessing only 5,000 out of these 20M features, which results in search times around 500 ms. The demonstration application is available online at http://disa.fi.muni.cz/demos/profiset-decaf/.

Acknowledgments. This work was supported by Czech Research Foundation project P103/12/G084.

Appendix I: Correctness of Algorithm 2

Lemma 1. *If d maintains Eq. (8) then Algorithm 2* GETNEXTIDS(q, j) *returns IDs of objects with the lowest j-th ranking $\psi_q^j(x)$, $j \in \Lambda$ that were not returned so far.*

Proof. The algorithm returns IDs from Q containing nodes and IDs from the j-th PPP-Tree. Because every node and ID is inserted into Q maximally once, the algorithm always returns something, unless all IDs were returned. Q is ordered by $d(q, \langle i_1, \ldots, i_{l'} \rangle)$ where $\langle i_1, \ldots, i_{l'} \rangle$ is either path to a node or it is equal to $\Pi_x^j(1..l)$ for ID$_x$ (recall that $d(q, \Pi_x^j(1..l))$ generate $\psi_q^j(x)$). Let ID$_x$ be returned by the algorithm; we prove the lemma by contradiction. Let us assume that there exists $y \in \mathcal{X}$: $d(q, \Pi_y^j(1..l)) < d(q, \Pi_x^j(1..l))$ and ID$_y$ was not returned by the algorithm. If ID$_y$ is in Q then it must be ahead of ID$_x$ (contradiction). Thus, ID$_y$ is not in Q, but Q must contain a node with path $\langle i_1, \ldots, i_{l'} \rangle$, $l' \leq l$ such that $\langle i_1, \ldots, i_{l'} \rangle = \Pi_y^j(1..l')$, because Q initially contains root of j-th PPP-Tree and then recursively all child nodes are inserted into Q (line 8). Because of (8), $d(q, \langle i_1, \ldots, i_{l'} \rangle) \leq d(q, \Pi_y^j(1..l)) < d(q, \Pi_x^j(1..l))$ which is in contradiction with the fact that ID$_x$ was on top of Q.

Appendix II: Optimizations of Algorithm 2

Complexity of the GETNEXTIDS routine is $O(|Q| \cdot \log |Q|)$ and the length of Q depends on "tightness" of Eq. (8). We propose the following optimizations for the d_Δ distance.

Optimization 1. Distance $d_\Delta(q, \Pi(1..l'))$ between $q \in \mathcal{D}$ and PP prefix on level $l' < l$ is corrected so that it returns the *minimum theoretical distance to PPP-Codes on level l with prefix $\Pi(1..l')$:*

$$d_\Delta'(q, \Pi(1..l')) = d_\Delta(q, \Pi(1..l')) \oplus \Pi_q(1..l-l')).$$

Notation $\oplus \Pi_q(1..l{-}l')$) is concatenation of the pivot indexes closest to the query. This addition does not break the condition (8) but, in our test cases, it resulted to reduction of the queue length by factor of 0.4–0.7.

Optimization 2. This optimization is relatively trivial: A leaf of the PPP-Tree at level $l' < l$ can keep IDs with the same PP suffix together as $\langle \Pi(l'{+}1..l); \mathrm{ID}_{x_1}, \ldots, \mathrm{ID}_{x_m} \rangle$ (see Sect. 4.1 for the original proposal); the list of IDs can be further optimized e.g. using delta encoding . This results in index memory reduction and, especially, in a slight reduction of the Q size, because such entry is inserted to Q only once.

Optimization 3. Another important cost component of the GETNEXTIDS algorithm are distances $d(q, \langle i_1, \ldots, i_{l'}, i_{l'+1} \rangle)$ evaluated for each item added into Q (lines 8 and 12). If the formula of distance d is a sum of independent values for each level from 1 to $l' + 1$ (as the d_Δ distance (4')) then value of $d(q, \langle i_1, \ldots, i_{l'}, i_{l'+1} \rangle)$ can be calculated as a sum of distance of its parent node $dist = d(q, \langle i_1, \ldots, i_{l'} \rangle)$ plus the addend for level $l' + 1$. In this case, the distances are calculated stepwise and no calculations are repeated.

References

1. Amato, G., Gennaro, C., Savino, P.: MI-File: using inverted files for scalable approximate similarity search. Multimedia Tools Appl. **71**(3), 1–30 (2012)
2. Amato, G., Savino, P.: Approximate similarity search in metric spaces using inverted files. In: Proceedings of InfoScale 2008. Vico Equense, Italy, June 4–6, pp. 1–10. ICST, Brussels, Belgium (2008)
3. Batko, M., Falchi, F., Lucchese, C., Novak, D., Perego, R., Rabitti, F., Sedmidubsky, J., Zezula, P.: Building a web-scale image similarity search system. Multimedia Tools Appl. **47**(3), 599–629 (2010)
4. Batko, M., Novak, D., Zezula, P.: MESSIF: metric similarity search implementation framework. In: Thanos, C., Borri, F., Candela, L. (eds.) Digital Libraries: Research and Development. LNCS, vol. 4877, pp. 1–10. Springer, Heidelberg (2007)
5. Beecks, C., Lokoč, J., Seidl, T., Skopal, T.: Indexing the signature quadratic form distance for efficient content-based multimedia retrieval. In: Proceedings of the ACM International Conference on Multimedia Retrieval, p. 8 (2011)
6. Bolettieri, P., Esuli, A., Falchi, F., Lucchese, C., Perego, R., Piccioli, T., Rabitti, F.: CoPhIR: A Test Collection for Content-Based Image Retrieval. CoRR, abs/0905.4 (2009)
7. Chávez, E., Figueroa, K., Navarro, G.: Effective proximity retrieval by ordering permutations. IEEE Trans. Patt. Anal. Mach. Intell. **30**(9), 1647–1658 (2008)
8. Christensen, D.: Fast algorithms for the calculation of Kendalls τ. Comput. Stat. **20**(1), 51–62 (2005)
9. Donahue, J., Jia, Y., Vinyals, O., Hoffman, J., Zhang, N., Tzeng, E., Darrell, T.: DeCAF: a deep convolutional activation feature for generic visual recognition. In: International Conference on Machine Learning, pp. 647–655 (2014)
10. Edsberg, O., Hetland, M.L.: Indexing inexact proximity search with distance regression in pivot space. In: Proceedings of SISAP 2010, Istanbul, Turkey, September 18–19, pp. 51–58. ACM Press, NY, USA (2010)

11. Esuli, A.: Use of permutation prefixes for efficient and scalable approximate similarity search. Inform. Process. Manag. **48**(5), 889–902 (2012)
12. Fagin, R., Kumar, R., Sivakumar, D.: Comparing top k lists. In: Proceedings of the Fourteenth Annual ACM-SIAM Symposium on Discrete Algorithms, SODA 2003, pp. 28–36. Society for Industrial and Appl. Math, Philadelphia, PA, USA (2003)
13. Fagin, R., Kumar, R., Sivakumar, D.: Efficient similarity search and classification via rank aggregation. In: Proceedings of ACM SIGMOD 2003. San Diego, California June 9–12, pp. 301–312. ACM Press, New York, USA (2003)
14. Gan, J., Feng, J., Fang, Q., Ng, W.: Locality-sensitive hashing scheme based on dynamic collision counting. In: Proceedings of the 2012 International Conference on Management of Data - SIGMOD 2012, pp. 541–552. ACM Press, New York, NY, USA (2012)
15. Gionis, A., Indyk, P., Motwani, R.: Similarity search in high dimensions via hashing. In: Proceedings of VLDB 1999, Edinburgh, Scotland, UK, September 7–10, pp. 518–529. Morgan Kaufmann (1999)
16. Jégou, H., Douze, M., Schmid, C.: Product quantization for nearest neighbor search. IEEE Trans. Patt. Anal. Mach. Intell. **33**(1), 117–128 (2011)
17. Krizhevsky, A., Sutskever, I., Hinton, G.E.: ImageNet classification with deep convolutional neural networks. In: Advances In Neural Information Processing Systems, pp. 1106–1114 (2012)
18. Muller-Molina, A.J., Shinohara, T.: Efficient similarity search by reducing I/O with compressed sketches. In: 2009 Second International Workshop on Similarity Search and Applications, pp. 30–38. IEEE, August 2009
19. Novak, D.: Multi-modal similarity retrieval with a shared distributed data store. In: Jung, J.J., Badica, C., Kiss, A. (eds.) INFOSCALE 2014. LNICST, vol. 139, pp. 28–37. Springer, Heidelberg (2015)
20. Novak, D., Batko, M., Zezula, P.: Metric Index: an efficient and scalable solution for precise and approximate similarity search. Inform. Syst. **36**(4), 721–733 (2011)
21. Novak, D., Batko, M., Zezula, P.: Large-scale Image retrieval using neural net descriptors. In: Proceedings of SIGIR 2015 (2015) (Will appear)
22. Novak, D., Kyselak, M., Zezula, P.: On locality-sensitive indexing in generic metric spaces. In: Proceedings of SISAP 2010, Istanbul, Turkey, September 18–19, pp. 59–66. ACM Press, New York, USA (2010)
23. Novak, D., Zezula, P.: Performance study of independent anchor spaces for similarity searching. Comput. J. **57**(11), 1741–1755 (2014)
24. Novak, D., Zezula, P.: Rank aggregation of candidate sets for efficient similarity search. In: Decker, H., Lhotská, L., Link, S., Spies, M., Wagner, R.R. (eds.) DEXA 2014, Part II. LNCS, vol. 8645, pp. 42–58. Springer, Heidelberg (2014)
25. Patella, M., Ciaccia, P.: Approximate similarity search: a multi-faceted problem. J. Discrete Algorithms **7**(1), 36–48 (2009)
26. Skala, M.: Counting distance permutations. J. Discrete Algorithms **7**(1), 49–61 (2009)
27. Torralba, A., Fergus, R., Weiss, Y.: Small codes and large image databases for recognition. In: 2008 IEEE Conference on Computer Vision and Pattern Recognition, pp. 1–8. IEEE, June 2008
28. Weiss, Y., Fergus, R., Torralba, A.: Multidimensional spectral hashing. In: Fitzgibbon, A., Lazebnik, S., Perona, P., Sato, Y., Schmid, C. (eds.) ECCV 2012, Part V. LNCS, vol. 7576, pp. 340–353. Springer, Heidelberg (2012)
29. Zezula, P., Amato, G., Dohnal, V., Batko, M.: Similarity Search: The Metric Space Approach. Advances in Database Systems, vol. 32. Springer, New York (2006)

Solving Data Mismatches in Bioinformatics Workflows by Generating Data Converters

Mouhamadou Ba[1]([✉]), Sébastien Ferré[2], and Mireille Ducassé[1]

[1] IRISA/INSA Rennes,
20 Avenue des Buttes de Coesmes, 35708 Rennes Cedex, France
mouhamadou.ba@irisa.fr
[2] IRISA/Université de Rennes 1,
263 Avenue Général Leclerc, 35042 Rennes Cedex, France

Abstract. Heterogeneity of data and data formats in bioinformatics entail mismatches between inputs and outputs of different services, making it difficult to compose them into workflows. To reduce those mismatches, bioinformatics platforms propose ad'hoc converters, called shims. When shims are written by hand, they are time-consuming to develop, and cannot anticipate all needs. When shims are automatically generated, they miss transformations, for example data composition from multiple parts, or parallel conversion of list elements.

This article proposes to systematically detect convertibility from output types to input types. Convertibility detection relies on a rule system based on abstract types, close to XML Schema. Types allow to abstract data while precisely accounting for their composite structure. Detection is accompanied by an automatic generation of converters between input and output XML data. We show the applicability of our approach by abstracting concrete bioinformatics types (e.g., complex biosequences) for a number of bioinformatics services (e.g., blast). We illustrate how our automatically generated converters help to resolve data mismatches when composing workflows. We conducted an experiment on bioinformatics services and datatypes, using an implementation of our approach, as well as a survey with domain experts. The detected convertibilities and produced converters were validated as relevant from a biological point of view. Furthermore the automatically produced graph of potentially compatible services exhibited a connectivity higher than with the ad'hoc approaches. Indeed, the experts discovered unknown possible connexions.

1 Introduction

Heterogeneity of data and data formats in bioinformatics entail mismatches between inputs and outputs of different services, making it difficult to compose them into workflows [1]. Formats to represent input and output data can be textual or based on XML technologies. Textual formats, often specific to a few services, have the advantage to be human readable [2]. XML formats despite verbosity are used for their expressiveness. To deal with these different formats in scientific workflows, special services, called *shims* are used for conversion of data

© Springer-Verlag Berlin Heidelberg 2016
A. Hameurlain et al. (Eds.): TLDKS XXIV, LNCS 9510, pp. 88–115, 2016.
DOI: 10.1007/978-3-662-49214-7_3

between services. Generally, they have to be manually defined (see, for example, Emboss [3], Galaxy [4], Mobyle [5]). Services for conversion of data represent more than 30 % of services in life science workflows according to the analysis of Wassink et al [6] on the Taverna Workflows. When composing services, users can get lost in specific parsers and shims required to transfer data between services. It is difficult to find appropriate shims because they are often mixed with other services. Users can be forced to create new shims, it is time-consuming and error prone. To avoid using shims, some developers implement their services so that they support several formats. It is a burden for service developers because they have to integrate several format conversions in each tool, and each format conversion may be duplicated across many services.

Formats based on XML technologies are proposed as standards to describe data types independently of tools [7,8]. BioXSD, for example, represents basic bioinformatics data [9]. It also allows meta-information to be added from ontologies, increasing the accuracy of representations. Yet, XML-based formats alone are not sufficient to solve the problem of data matching. On the one hand, most formats are textual, thus it is important to be able to match services using XML *and* textual formats. On the other hand, even if all formats were XML standardized, it remains to solve the n:m matching problem [10], namely matching and conversion between two *composite structures*, i.e. XML trees.

Work related to data mismatches in scientific workflows addresses, among others, classifying mismatches, matching and resolving mismatches. For example, Li et al. [11] classify service composition mismatches. The classification of mismatches enables to understand problems and to find appropriate solutions. Besides approaches that address verifying matching between services (e.g., Lebreton al. [12]), there are approaches that resolve data mismatches by means of shims to insert between services. Among them, some approaches search shims in existing libraries [13,14] and other approaches automatically generate shims [15,16]. Many approaches use ontologies to verify matching between service parameters. However, they do not generally guarantee parameter compatibility at syntactic level. Approaches that find shims, for example Elizondo et al. [13] and Hull et al. [14], fix data mismatches but expect the shims to be provided by third-parties. Approaches that generate shims automatically provide data transformers to insert between services. For example, Kaslev et al. [16] generate transformations at workflow execution time. Dibernado et al [17] provide possible transformations during workflow construction. However, these approaches miss transformations such as data composition from multiple parts, or parallel conversion of list elements.

This article proposes a new approach to generate shims to help compose services.[1] It supports complex data representations and transformations. It systematically detects convertibility from output types to input types. Convertibility detection relies on abstract types, close to XML Schema, abstracting data while precisely accounting for their composite structure. The main contribution is the definition of convertibility rules that exploit composition and decomposition as

---

[1] It is a revised and extended version of Ba et al. [18].

well as specialization and generalization of types. Furthermore, the rules automatically generate a complete constructive specification of the conversion from output to input types. That specification enables to generate executable converters between input and output XML data. We report an experiment on bioinformatics services with an implementation of our approach. We manually specified the inputs and outputs of services using abstract types, where each service is understood as a function from input(s) to output(s) as proposed by Missier et al [19]. The detected convertibilities and produced converters were analyzed with a team of the GenOuest[2] bioinformatics platform. They have been reckoned relevant from a biological point of view. We also led a survey with domain experts, that shows the relevance of connections established between bioinformatics services. When adding a new service, with our approach, it is sufficient to define or reuse abstract types for input and output data, and the new service will be automatically integrated in the global system. At present, in order to achieve the same goal, many converters have to be manually developed for services which are not immediately compatible. That is significantly heavier than specifying a few abstract types. Furthermore, identifying the compatible services by hand is already a challenge while our approach does it automatically. As a consequence, our approach automatically produces a graph of potentially compatible services with a connectivity higher than with the ad'hoc approaches.

In the following, Sect. 2 introduces the abstract representation of types. Section 3 defines convertibility between abstract types by a rule system, and proves that it forms a reflexive and transitive relationship. Section 4 shows how to instantiate our method and provides a use case in bioinformatics. Section 5 presents an experiment in bioinformatics, and a survey with domain experts. Section 6 compares our approach with related work. Section 7 gives some perspectives.

2   Representation of Types

In this section we present the language used to describe the types of data. It is defined from an open set of primitives and a fixed set of type constructors. From a semantic point of view, a type denotes a set of XML values. An XML value is a sequence of XML nodes. An XML sequence may be empty or contain a single node. An XML node is either an XML element or a textual element (CDATA). An XML element is made of a tag and a content, which is a sequence of XML nodes. Our language of types follows the main XML Schema constructs, but to simplify the presentation we use regular expressions inspired by the work of Hosoya et al. [20]. In the following, types are denoted by uppercase letters (e.g., T, T_1) and function $XML(T)$ defines the semantics of type T by a set of XML values.

- **Primitive types:** $T = p$, where p is a primitive type. Primitive types are the basic ingredients to build other types. Their structure is atomic, they are

---

[2] http://www.genouest.org/.

not decomposable. XML instances of a primitive type are CDATA (text). For example, *int* is a primitive type representing integers.

– **Constructor *tag*:** $t[T_1]$ where t is a tag. This expression denotes XML elements whose tag is t and whose content is of type $T_1 : XML(t[T_1]) = \{<t> x_1 </t> \mid x_1 \in XML(T_1)\}$. Tags provide semantics for data and can be bound to concepts of ontologies. In XML Schema, constructor *tag* can be expressed by a tag or by the attribute *name* of tag *xs:element*.

– **Constructor *empty*:** ε. The empty XML sequence: $XML(\varepsilon) = \{\varepsilon\}$.

– **Constructor *tuple*:** $T_1 T_2$. This type expression denotes XML sequences that are the concatenation of instances of T_1 and instances of $T_2 : XML(T_1 T_2) = \{x_1 x_2 \mid x_1 \in XML(T_1), x_2 \in XML(T_2)\}$. That constructor is used to define composite types and sequences. In XML Schema, constructor *tuple* can be expressed by tags *xs:complexType* and *xs:sequence*.

– **Constructor *union*:** $T_1 | T_2$. This type expression denotes the union of instances of T_1 and instances of $T_2 : XML(T_1 | T_2) = XML(T_1) \cup XML(T_2)$. Constructor *union* can be used, for example, to consider several different types and make some treatments without distinction. In XML Schema, that constructor can be expressed by tag *xs:choice*.

– **Constructor *list*:** T_1+. This type expression denotes the non-empty sequences of instances of type T_1 (homogeneous lists): $XML(T_1+) = \{x_1 \ldots x_n \mid n \geq 1 \wedge x_1, \ldots, x_n \in T_1\}$. In XML Schema, that constructor can be expressed by the *minOccurs* and *maxOccurs* attributes associated with tag *xs:sequence*.

– **Constructor *optional*:** $T_1?$. This type expression is equivalent to $T_1 | \varepsilon$.

For abbreviation it is possible to bind a name to a type expression. For example, in the expression $T = T_1 T_2$, there is a type name T allowing to simplify the re-use of type expressions, where the name T can be used to refer to $T_1 T_2$. We do not allow recursive type expressions yet.

3   Convertibility Rules

By using type theory in the context of workflows, we follow the same line as previous work on web services such as the one of Chen et al. [21]. The novelty of our approach lies in the definition of a rule system to prove convertibility between types, and to apply the proof-as-program paradigm [22] to automatically derive executable converters from convertibility proofs.

In order to motivate the following convertibility rules, we list a few typical cases where there is a mismatch between two types A and B ($A \neq B$), whereas there is a semantic match, i.e. A can be converted to B.

– A and B use tags that are different but have the same meaning, for example `Integer` and `Int`.
– B may be replaced by A, for example `Float` by `Long`.
– B is a concatenation of some components of A, for example `person[name tel email]` from `person[name contact[address tel email]]`.
– A is subsumed by B, for example `protein seq` by `biological seq`.

$$\frac{f_p : p_1 \rightarrow_p p_2}{f : p_1 \rightarrow p_2 \quad f(x) = f_p(x)} \qquad \text{(PRIMITIVE)}$$

$$\frac{t_a \rightarrow_t t_b \quad f_1 : A \rightarrow B}{f : t_a[A] \rightarrow t_b[B] \quad f(x) = element(t_b, f_1(content(x)))} \qquad \text{(TAGCHANGE)}$$

$$\frac{f_1 : A \rightarrow B}{f : t[A] \rightarrow B \quad f(x) = f_1(content(x))} \qquad \text{(TAGREMOVAL)}$$

$$\frac{}{f : A \rightarrow \varepsilon \quad f(x) = \varepsilon} \qquad \text{(EMPTY)}$$

$$\frac{f_1 : A \rightarrow B_1 \quad f_2 : A \rightarrow B_2}{f : A \rightarrow B_1 B_2 \quad f(x) = concat(f_1(x), f_2(x))} \qquad \text{(CONCAT)}$$

$$\frac{f_1 : A_1 \rightarrow B}{f : A_1 A_2 \rightarrow B \quad f(x) = (let\ x_1, x_2 = select(x, A_1, A_2)\ in\ f_1(x_1))} \qquad \text{(LEFTSELECTION)}$$

$$\frac{f_2 : A_2 \rightarrow B}{f : A_1 A_2 \rightarrow B \quad f(x) = (let\ x_1, x_2 = select(x, A_1, A_2)\ in\ f_2(x_2))} \qquad \text{(RIGHTSELECTION)}$$

$$\frac{f_1 : A_1 \rightarrow B \quad f_2 : A_2 \rightarrow B}{f : A_1 | A_2 \rightarrow B \quad f(x) = (case\ (x : A_1)\ then\ f_1(x)\ |\ (x : A_2)\ then\ f_2(x))} \qquad \text{(PRECHOICE)}$$

$$\frac{f_1 : A \rightarrow B_1}{f : A \rightarrow B_1 | B_2 \quad f(x) = f_1(x)} \qquad \text{(LEFTPOSTCHOICE)}$$

$$\frac{f_2 : A \rightarrow B_2}{f : A \rightarrow B_1 | B_2 \quad f(x) = f_2(x)} \qquad \text{(RIGHTPOSTCHOICE)}$$

$$\frac{f_1 : A \rightarrow B}{f : A+ \rightarrow B+ \quad f(x) = map(f_1, x)} \qquad \text{(MAP)}$$

$$\frac{f_1 : A \rightarrow B}{f : A+ \rightarrow B \quad f(x) = (let\ x_1 = choose(x, A)\ in\ f_1(x_1))} \qquad \text{(CHOICE)}$$

$$\frac{f_1 : A \rightarrow B}{f : A \rightarrow B+ \quad f(x) = f_1(x)} \qquad \text{(SINGLETON)}$$

Fig. 1. Convertibility rules and definitions of generated converters. For example, rule TagChange reads as follows: if tag t_a is convertible to tag t_b and if f_1 is a converter from A to B, then there is a converter f from $t_a[A]$ to $t_b[B]$ such that applying f to a data x involves extracting its content, applying f_1 to the content, then re-encapsulate the result with tag t_b.

3.1   Rule System

Figure 1 lists all the rules that specify when a type A is convertible to a type B. They also define associated converters as functions from A to B. Those rules form a natural deduction system whose judgements are in the form $f : A \rightarrow B$,

Table 1. Utility functions on XML values.

Function	Input	Output	Description
Content	XML	XML	Returns the content of an XML element
Element	Tag, XML	XML	Builds an XML element given a tag and an XML content
Concat	XML, XML	XML	Returns the concatenation of two XML sequences
Select	XML, type, type	XML, XML	Splits an XML sequence in two parts matching given types
Map	Converter, XML	XML	Applies a converter to each node of an XML sequence and returns the concatenation of the results
Choose	XML, type	XML	Returns any element matching a given type from an XML sequence

i.e. f is a converter from an XML value of type A to an XML value of type B, and hence A is convertible to B. A judgement $f : A \to B$ holds true if and only if it is possible to build a proof tree with that judgement at the root, and where each node instantiates a rule. The deduction system works by structural induction on couples of types (A, B), covering all combinations of type constructors for which convertibility is possible. The rules depend on conversion axioms for tags ($t_a \to_t t_b$), and on converters between primitive types ($f_p : p_1 \to_p p_2$). Those base conversions depend on the application domain, and correspond, for instance, to well-known conversion functions (e.g., from floats to integers). By default, we assume that $t \to_t t$ for every tag t, and $f_p : p \to_p p$ with $f_p(x) = x$ for every primitive type p. The definitions of converters in rules make use of utility functions on XML values, which are described in Table 1. Figure 2 shows a convertibility proof.

Primitive Rule. Rule (PRIMITIVE) allows the use of a primitive converter f_p when the two types are primitive types. That rule handles the conversion of the leaves of XML trees (CDATA nodes).

Tag Rules. These rules handle the conversion from and to XML elements. Rule (TAGCHANGE) defines converters from an XML element x to another XML element $element(t_b, f_1(content(x)))$ by applying a domain-dependent tag conversion (here, from t_a to t_b), and by recursively applying a converter f_1 to the content of x. Function *content* gives access to the content of an XML element, and function *element* builds the new element from the converted tag and converted content. Rule (TAGREMOVAL) define converters from an XML element to an XML sequence by ignoring the tag, and recursively converting the content.

Empty and Tuple Rules. These rules handle conversions of XML sequences, i.e. constructors *empty* and *tuple*. Rule (EMPTY) says that any XML value x can be converted to the empty XML sequence ϵ. Rule (CONCAT) defines converters that first apply the converters f_1 and f_2 to the source value x, and then concatenates the two results $f_1(x)$ and $f_2(x)$ with function *concat*, hence producing an XML sequence. Rules (LEFTSELECTION) and (RIGHTSELECTION) define converters that select respectively the left and right part of the source data, an XML sequence, and convert it to the target data. This is useful when only a part of the source data is necessary to produce the target data. The selection of the parts (function *select*) is guided by the sub-types of the source sequence.

Union Rules. These rules handle conversions from and to unions of types. Rule (PRECHOICE) defines converters that produce a target data using a different converter depending on the type of source data (A_1 or A_2). This is useful when the source data can have different structures (union type). Rules (LEFTPOSTCHOICE) and (RIGHTPOSTCHOICE) choose a converter to a target subtype, when the target type is an union. This is useful when the target data has several acceptable structures.

List Rules. The remaining rules handle conversions from and to lists. Rule (MAP) define converters from a source list to a target list where a same converter is applied to each element of the list. Function *map* is used to perform iteration over list elements, and concatenation of converted elements. Rule (CHOICE) defines converters that first choose an element of a list, and then recursively apply a converter to it. This is useful when a single element is expected while a list is provided. Rule (SINGLETON) defines converters that produce singleton lists from a source element, after recursively applying a converter to it. This is useful when a list is expected while a single element is provided.

For a given couple (A, B) of type expressions, several rules may be applicable. In that case, it is sufficient that one of them leads to a success to prove the convertibility from A to B. Figure 2 details the proof of convertibility between two kinds of biological sequence lists. In the source list, sequences are made of a nucleotide sequence, a species name, and a version number, while in the target list, sequences are made of an organism, and a nucleotide sequence. Another difference is that nucleotide sequences are lowercase in the source list (primitive type *acgt*), and uppercase in the target list (primitive type *ACGT*). In the proof (Fig. 2), we assume that a primitive converter is available to convert from lowercase to uppercase (see step 1.2.2.1.2.), and domain knowledge tells us that tag *species* can be replaced by *organism* (see step 1.2.1.1.1.1.). Each item in Fig. 2 is the conclusion of a rule, and the sub-items are the hypotheses of the rule. At each item, the converter function is defined with calls to the converter function of sub-items. After inlining the definition of intermediate functions in the main function f, we obtain the full definition of f in Fig. 3.

Our rule system exhibits two kinds of non-determinism: (1) in the generation of converters, and (2) in the definition of converters. Firstly, given two types A and B, the system may generate several converters from A to B, i.e. several solutions to the conversion problem. This is a common feature of rule systems.

seq[ns[acgt] species[string] version[int]]+ → seq[organism[string] ns[ACGT]]+
(MAP) $f(x) = map(f_1, x)$
   1. seq[ns[acgt] species[string] version[int]] → seq[organism[string] ns[ACGT]]
   (TAGCHANGE) $f_1(x) = element(seq, f_{1.2}(content(x)))$
     1.1. seq $\rightarrow_t$ seq
     1.2. ns[acgt] species[string] version[int] → organism[string] ns[ACGT]
     (CONCAT) $f_{1.2}(x) = concat(f_{1.2.1}(x), f_{1.2.2}(x))$
       1.2.1. ns[acgt] species[string] version[int] → organism[string]
       (RIGHTSELECTION) $f_{1.2.1}(x) = (let\ x_1, x_2 = select(x)\ in\ f_{1.2.1.1}(x_2))$
         1.2.1.1. species[string] version[int] → organism[string]
         (LEFTSELECTION) $f_{1.2.1.1}(x) = (let\ x_1, x_2 = select(x)\ in\ f_{1.2.1.1.1}(x))$
           1.2.1.1.1. species[string] → organism[string]
           (TAGCHANGE) $f_{1.2.1.1.1}(x) = element(organism, f_{1.2.1.1.1.2}(content(x)))$
             1.2.1.1.1.1. species $\rightarrow_t$ organism
             1.2.1.1.1.2. string $\rightarrow_p$ string
             (PRIMITIVE) $f_{1.2.1.1.1.2}(x) = x$
       1.2.2. ns[acgt] species[string] version[int] → ns[ACGT]
       (LEFTSELECTION) $f_{1.2.2}(x) = (let\ x_1, x_2 = select(x)\ in\ f_{1.2.2.1}(x_1))$
         1.2.2.1. ns[acgt] → ns[ACGT]
         (TAGCHANGE) $f_{1.2.2.1}(x) = element(ns, f_{1.2.2.1.2}(content(x)))$
           1.2.2.1.2. acgt $\rightarrow_p$ ACGT
           (PRIMITIVE) $f_{1.2.2.1.2}(x) = uppercase(x)$

Fig. 2. An example proof tree of convertibility between two kinds of sequence lists.

  $f(x) = map(f_1, x)$
  where $f_1(x) = element(seq, concat($
      $let\ x_1, x_2 = select(content(x), ns[acgt], (species[string]\ version[int]))$
       $in\ let\ x_{21}, x_{22} = select(x_2, species[string], version[int])$
         $in\ element(organism, content(x_{21})),$
      $let\ x_1, x_2 = select(content(x), ns[acgt], (species[string]\ version[int]))$
       $in\ element(ns, uppercase(content(x_1)))))$

Fig. 3. The generated converter for example of Fig. 2

For example, a converter $f : AA \rightarrow A$ can be produced by either Rule (LEFT-SELECTION) or Rule (RIGHTSELECTION): in the former, the left part of the source value is selected, while in the latter, the right part is selected. In practice, one converter must be chosen, which could be done through user interaction. Secondly, a generated converter may produce several target values for a same source value. This non-determinism comes from some utility functions, and the *case* construct. Function *select* may find different ways to split the source value in two parts. Function *choose* has as many results as elements in the list. The *case* construct has two results when the two conditions are satisfied, when x matches both types A_1 and A_2. This second form of non-determinism could be used to express iteration in a workflow. For example, assuming that service S_1 produces lists of sequences, and service S_2 consumes one sequence at a time, the converter generated by Rule (CHOICE) could be a way to express that S_2 must be iterated over the results of S_1, and the output of S_2 could be considered to be the list of individual results. It will meet needs for job iteration in bioinformatics workflows [23].

3.2   Properties: Reflexivity and Transitivity

Two important properties of the convertibility relationship are *reflexivity* and *transitivity*. First, every type A is convertible to itself and it suffices to take the identity function as a converter. Second, for any types A, B, C, if A is convertible to B, and B is convertible to C, then A is convertible to C and it suffices to compose the two converters from A to B and from B to C to obtain a converter from A to C. We formalize those two properties in the following theorems, and give their proofs. In those theorems, we assume that tag convertibility ($\rightarrow_t$) and primitive convertibility ($\rightarrow_p$) are reflexive and transitive. As a consequence, unlike the approach of Kaslev et al. [16], we do not need rules for transitivity and reflexivity in our rule system. This makes convertibility proofs simpler and more efficient.

Theorem 1. *Let A be a type expression. There is a proof in the rule system of the judgement $f : A \rightarrow A$ where $f(x) = x$.*

Proof. We proceed by induction on type A, considering the six type constructors as different cases. For each of the type constructor, the property is verified because:

1. $A = p$ (primitive): from assumption on primitives ($p \rightarrow_p p$), and by applying Rule (PRIMITIVE).
2. $A = t[A_1]$ (tag): from induction hypothesis on A_1 ($A_1 \rightarrow A_1$), and assumption on tags ($t \rightarrow_t t$), and by applying Rule (TAGCHANGE).
3. $A = \epsilon$ (empty): from Rule (EMPTY).
4. $A = A_1 A_2$ (tuple): from induction hypothesis on A_1 ($A_1 \rightarrow A_1$) and A_2 ($A_2 \rightarrow A_2$), by applying Rule (LEFTSELECTION) to the first, and Rule (RIGHTSELECTION) to the second, and finally by applying Rule (CONCAT) to the consequences of the two previous rules.
5. $A = A_1 | A_2$ (union): from induction hypothesis on A_1 and A_2, by applying Rule (LEFTPOSTCHOICE) to the first (introducing A_2), and Rule (RIGHTPOSTCHOICE) to the second (introducing A_1), and finally by applying Rule (PRECHOICE) to the consequences of the two previous rules.
6. $A = A_1 +$ (list): from induction hypothesis on A_1, and by applying Rule (MAP).

In each case, it can be shown that the produced converter function is equivalent to the identity function. For instance, in the *tag* case, assuming $f_1 : A_1 \rightarrow A_1$ is equivalent to the identity function (induction hypothesis), it can be shown that the resulting function $f(x) = element(t, f_1(content(x)))$ is equivalent to $element(t, content(x))$ which is equal to x because x has type $t[A_1]$. ∎

Theorem 2. *Let A, B, C be expression types. If there are proofs in the rule system of the judgements $f_1 : A \to B$ and $f_2 : B \to C$, then there is also a proof of the judgement $f : A \to C$ where $f(x) = f_2(f_1(x))$.*

Proof. We proceed by induction on the rules that are used at the root of the proofs of $A \to B$ and $B \to C$. As there are 13 distinct rules, there are potentially 169 distinct cases to consider. Fortunately, many cases have similar proofs and can be grouped together based on the distinction between three kinds of rules:

- Constructors ($\mathcal{C}$): rules where only the target type changes between premises and conclusion (Rules (CONCAT), (LEFTPOSTCHOICE), (RIGHT-POSTCHOICE), (SINGLETON), (EMPTY)),
- Destructors ($\mathcal{D}$): rules where only the source type changes between premises and conclusion (Rules (TAGREMOVAL), (LEFTSELECTION), (RIGHTSELECTION), (PRECHOICE), (CHOICE)),
- Transformers ($\mathcal{T}$): rules where both source and target change but use the same kind of type (Rules (PRIMITIVE), (TAGCHANGE), (MAP)).

Using that grouping, we arrive at 6 meta-cases described using the above group codes $\mathcal{C}, \mathcal{D}, \mathcal{T}$ and $\mathcal{X}$ to mean any rule. Applying unification constraints on the middle type B, those meta-cases then decompose themselves in 21 elementary cases:

1. $\mathcal{X} - \mathcal{C}$:
   (a) $\mathcal{X}$ - (CONCAT): the proof of $B \to C$ by Rule (CONCAT) implies that $C = C_1 C_2$, and that we have proofs for $B \to C_1$, and $B \to C_2$. By induction hypothesis on A, B, C_1 and A, B, C_2, we obtain $A \to C_1$ and $A \to C_2$. Then, by applying Rule (CONCAT) on those judgements, we finally obtain $A \to C$.
   (b) $\mathcal{X}$ - (LEFTPOSTCHOICE): we have $C = C_1 | C_2$, and $B \to C_1$. By induction hypothesis on A, B, C_1, we obtain $A \to C_1$. Then, by applying Rule (LEFTPOSTCHOICE) on the later, we obtain $A \to C_1 | C_2$.
   (c) $\mathcal{X}$ - (RIGHTPOSTCHOICE): similar to previous case.
   (d) $\mathcal{X}$ - (SINGLETON): we have $C = C_1 +$ and $B \to C_1$. By induction hypothesis on A, B, C_1, we obtain $A \to C_1$, from which we obtain $A \to C$ by applying Rule (SINGLETON).
   (e) $\mathcal{X}$ - (EMPTY): we have $C = \epsilon$. We directly obtain $A \to C$ by applying Rule (EMPTY) (everything is convertible to ϵ).
2. $\mathcal{D} - \mathcal{X}$:
   (a) (TAGREMOVAL) - $\mathcal{X}$: we have $A = t[A_1]$ and $A_1 \to B$. By induction hypothesis on A_1, B, C, we obtain $A_1 \to C$. By applying Rule (TAGRE-MOVAL) on the latter, we obtain $A \to C$.
   (b) (LEFTSELECTION) - $\mathcal{X}$: we have $A = A_1 A_2$, and $A_1 \to B$. By induction hypothesis on A_1, B, C, we obtain $A_1 \to C$. By applying Rule (LEFTSE-LECTION) on the latter, we obtain $A \to C$.
   (c) (RIGHTSELECTION) - $\mathcal{X}$: similar to previous case.

(d) (PRECHOICE) - $\mathcal{X}$: we have $A = A_1|A_2$, $A_1 \to B$, and $A_2 \to B$. By induction hypothesis on A_1, B, C and A_2, B, C, we obtain $A_1 \to B$ and $A_2 \to B$. By applying Rule (PRECHOICE), we obtain $A \to C$.

(e) (CHOICE) - $\mathcal{X}$: we have $A = A_1+$ and $A_1 \to B$. By induction hypothesis on A_1, B, C, we obtain $A_1 \to C$. By applying Rule (CHOICE) to the latter, we obtain $A \to C$.

3. $\mathcal{C}$ - $\mathcal{D}$:

(a) (CONCAT) - (LEFTSELECTION): we have $B = B_1 B_2$, and the judgements (1) $A \to B_1$, (2) $A \to B_2$, (3) $B_1 \to C$. By induction hypothesis on A, B_1, C and judgements (1) and (3), we obtain $A \to C$.

(b) (CONCAT) - (RIGHTSELECTION): similar as previous case.

(c) (LEFTPOSTCHOICE) - (PRECHOICE): we have $B = B_1|B_2$ and the judgements $A \to B_1$, $B_1 \to C$, and $B_2 \to C$. By induction hypothesis on A, B_1, C, we obtain $A \to C$.

(d) (RIGHTPOSTCHOICE) - (PRECHOICE): similar to previous case.

(e) (SINGLETON) - (CHOICE): we have $B = B_1+$, and the judgements $A \to B_1$ and $B_1 \to C$. By induction hypothesis on A, B_1, C, we obtain $A \to C$.

4. $\mathcal{T}$ - $\mathcal{T}$:

(a) (PRIMITIVE) - (PRIMITIVE): we have $A = p_1$, $B = p_2$, $C = p_3$, and $p_1 \to_p p2$ and $p_2 \to_p p_3$. From assumptions on primitives, we obtain $p_1 \to_p p_3$. By applying Rule (PRIMITIVE) on the latter, we obtain $A \to C$.

(b) (TAGCHANGE) - (TAGCHANGE): we have $A = t_A[A_1]$, $B = t_B[B_1]$, $C = t_C[C_1]$, and $t_A \to_t t_B$, $t_B \to_t t_C$, $A_1 \to B_1$, $B_1 \to C_1$. From assumptions on tags, we obtain $t_A \to_t t_C$. By induction hypothesis on A_1, B_1, C_1, we obtain $A_1 \to C_1$. By applying Rule (TAGCHANGE) to the two latter judgements, we obtain $A \to C$.

(c) (MAP) - (MAP): we have $A = A_1+$, $B = B_1+$, $C = C_1+$, and $A_1 \to B_1$, $B_1 \to C_1$. By induction hypothesis on A_1, B_1, C_1, we obtain $A_1 \to C_1$. By applying Rule (MAP) to the latter, we obtain $A \to C$.

5. $\mathcal{C}$ - $\mathcal{T}$:

(a) (SINGLETON) - (MAP): we have $B = B_1+$, $C = C_1+$, and $A \to B_1$, $B_1 \to C_1$. By induction hypothesis on A, B_1, C_1, we obtain $A \to C_1$. By applying Rule (SINGLETON) to the latter, we obtain $A \to C$.

6. $\mathcal{T}$ - $\mathcal{D}$:

(a) (TAGCHANGE) - (TAGREMOVAL): we have $A = t_A[A_1]$, $B = t_B[B_1]$, and $A_1 \to B_1$, $B_1 \to C$. By induction hypothesis on A_1, B_1, C, we obtain $A_1 \to C$. By applying Rule (TAGREMOVAL) to the latter, we obtain $A \to C$.

(b) (MAP) - (CHOICE): we have $A = A_1+$, $B = B_1+$, and $A_1 \to B_1$, $B_1 \to C$. By induction hypothesis on A_1, B_1, C, we obtain $A_1 \to C$. By applying Rule (CHOICE) to the latter, we obtain $A \to C$.

In each case, it can be shown that the produced converter function is equivalent to the composition of the two converters from A to B, and from B to C. For instance, in the $\mathcal{X}$ - (CONCAT) case, assuming $f_1 : A \to B$, and $f_2 : B \to C$, we have $f_2(x) = concat(f_2'(x),\ f_2''(x))$ where $f_2' : B \to C_1$ and $f_2'' : B \to C_2$.

By application of three rules as indicated in the proof, we obtain for $f : A \rightarrow C$, the definition $f(x) = concat(f'_2(f_1(x)), \ f''_2(f_1(x)))$. From the definition of f_2, that definition can be simplified into $f(x) = f_2(f_1(x))$, which is indeed the composition of f_1 and f_2. ∎

3.3 Implementation

We implemented our rule system in a program that decides the convertibility between any two type expressions, and generates converters from data matching the first type expression to data matching the second type expression. The algorithm is directly derived from the above rules and combines *pattern matching* on type expressions to identify constructors, and recursive calls on type subexpressions. The examination of rules shows that recursive calls always involve smaller couples of expressions, which ensures termination of the program in all cases. The computation time required to decide convertibility may be important when the input type is very large, because of the non-deterministic nature of the rule system. However, convertibility is computed once for a set of types. In practice, types are not very large, and we have not encountered any difficulty in our experiments to compute all convertibilities for a set of bioinformatic services (see Sect. 5.1). Generated converters are efficient. Most rules imply a constant cost per XML node, and hence a linear complexity over the input data. The two cases that may imply additional costs concern node duplication (see Rules LEFT-SELECTION and RIGHTSELECTION) and choice handling (see Rule PRECHOICE). Node duplication corresponds to the situation where a node of the input data is converted to several nodes of the output data. In this case, the duplicated node is processed several times, thus exceeding linear complexity. However, the number of duplications is bounded by the number of constructors in the output type. Choice handling corresponds to the situation where a node can have one of two types (A_1 or A_2), and the correct type has to be identified by the converter at execution time. Type identification requires one additional node processing for each choice, thus exceeding linear complexity. However, the number of choices is equal to the number of constructor *union* in the input type, and the additional processing may only concern a fragment of the input data. In practice, input and output types of bioinformatics services make a limited use of duplications and choices, thus in the worst case, the complexity of converters is in the size of the input data multiplied by a small constant. Our generated converters are represented in XQuery, a suitable language to process XML documents, which makes the converters executable. To account for non-determinism, the result of our program is a collection of converters. Each converter will be a function from an XML value to an XML value. In the case of non-deterministic converters, only one value is produced so far. The production of several values is left to future implementation.

4   Instantiation and Use Case

This section presents how we instantiate our type abstraction to bioinformatics data. It also presents a use case that shows how our approach can be used to detect and resolve data mismatches in bioinformatics workflows.

```
> Accession = accession[string]
> SSeq = simpleSequence[string]
> ProtSeq = ns[sSeq]
> DNASeq = as[sSeq]
> Bioseq =  DNASeq | ProtSeq
> CBioseq = complexBiosequence[
        seq[Bioseq]
        species[string] source[string] name[string]
        version[string] note[string]?]
> CProtSeq = complexProteinSequence[
        seq[ProtSeq]
        species[string] source[string] name[string]
        version[string] note[string]?]
> BioseqList = CBioseq+
```

Fig. 4. Examples of bioinformatics types

4.1   Instantiation to Bioinformatics

Depending on application requirements and on the nature of the data in bioinformatics platforms, many formats are available. To represent genomics data, various textual formats (e.g., FastQ, BED)[3] and XML formats (e.g., BioXSD[4], phyloXML[5]) are provided. Textual formats are the most commonly used. In addition to data formats, ontologies, such as EDAM[6] have been proposed to organize and classify resources including data types and formats. Our work starts from these resources to define input and output types of services. We abstract data types by focusing on the information contents and composite structure of data. Figure 4 shows simple examples of types defined manually from existing bioinformatic formats. *Accession* and *SSeq* are simple types representing, respectively, an accession number and a raw sequence, defined with a constructor tag and a primitive. In the same way, *DNASeq* (representing nucleotide sequences) and *ProtSeq* (representing amino acid sequences) are defined using *SSeq*, they specialize the sequences. Their union forms *Bioseq*, a biological sequence generalizing the sequences. *CBioseq* and *CProtSeq* are composite types holding several types through constructor *tuple*. They are biological sequences containing required (e.g., *sequence[Bioseq]*) and optional (e.g.,

---

[3] http://genome.ucsc.edu/FAQ/FAQformat.html/.
[4] http://bioxsd.org/.
[5] http://www.phyloxml.org/.
[6] http://edamontology.org.

note[*string*]?) contents, *CProtSeq* being more specific than *CBioseq*. *CBioseqList* defines a list of *CBioseq* using constructor *list*. The other types we use are defined in the same way as the above types. Labels are inspired from the EDAM ontology.

Compared to data types and formats used on platform EMBOSS[7], our types can define accession numbers allowing to represent, for example, sequence and database references. They can represent raw sequences as in plain text format, single sequences as in gcc format, one or several sequences (e.g., alignment of sequences) as in FASTA format, as well as a simple sequence associated to its annotations and features as in EMBL format. Our types can also represent lists of files and differentiate the nature of information contained in files, for example, nucleotide sequence versus amino acid sequence. We take into account data types and formats commonly used for inputs and outputs of services on platform EMBOSS. Most platforms we visited use the same categories of data types and formats. Compared to XML formats such as BioXSD, our abstraction represents contents at a higher abstraction level. We only consider information relevant for our matching between input and output data of services. We ignore, for example some type attributes and type restrictions irrelevent for current input and output data used in services. If necessary, they can easily be added.

Abstraction of types is straightforward for XML formats thanks to XML schemas. For textual formats, informal specifications must be studied to derive a structural representation. Our experiment with genomics types shows that type expressions recur frequently, they can easily be reused after being defined once. Since the most common data types are defined, there are increasingly less types to define. The BioXSD initiative defines several data types for common bioinformatics web services. Specialized XML formats, such as phyloXML [8] for phylogenetic data and PDBML [24] for systems biology, exist for sub-domains of bioinformatics. Moreover, XML alternatives are provided for some textual formats (e.g., GFF [25]) and some platforms define their own XML format (e.g., Uniprot XML [26]). For our experiment, the abstraction of types is done manually but the spreading of the above mentioned solutions will facilitate the task. We can even expect automatic or semi-automatic abstraction processes.

The defined types represent inputs and outputs of current genomics services. In our approach, adding a new service requires two steps. Firstly, identify the abstract types used as inputs and outputs of the service. Secondly, implement, if they do not already exist, the converters between XML schemas and each format, since our types define an XML. Unlike other approaches, it is not necessary to define converters for all pairs of formats, but only two for each format (from and to XML).

4.2 Use Case: Resolving Data Mismatches in a Bioinformatics Workflow

We now present a workflow (*w*), constructed with our approach. It compares a consensus sequence produced from an alignment of a list of sequences with

---

[7] http://emboss.sourceforge.net/docs/Themes.

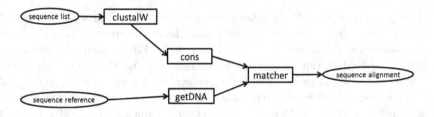

Fig. 5. Workflow at user level (w_u).

another sequence obtained from an accession number. The workflow consumes sequence lists (tabular form) and references (accessions) to DNA sequences. It produces sequence alignments. In the following, we describe three representations of workflow w, at three different levels. The workflow at user level is what the user expects to see but we show that it is underspecified, and cannot be executed as such. The workflow at execution level is fully specified and executable, but it is over-detailed for the user because it confuses genuine services and shims. We finally introduce the workflow at a abstract level, from which both user level and execution level representations can be derived automatically.

Figure 5 shows the workflow at a user level (w_u). Boxes represent tasks and ellipses represent input and output data of the workflow. It is an interconnection of inputs and outputs of services from bioinformatics platforms. We assume that a service performs one task, it may have one or several inputs and one or several outputs. To simplify, we do not take into account parameters used to manage service behaviour (e.g., algorithm parameters). The workflow uses the following services:

- **matcher** compares two biological sequences. It takes as inputs two biological sequences in fasta files and returns as output a MSF alignment.
- **getDNA** retrieves a DNA sequence from a database. It takes an accession number and returns an embl file containing a DNA sequence.
- **cons** creates a consensus sequence from a multiple alignment. It takes a MSF alignment and returns a fasta containing a biological sequence.
- **clustalW** makes a multi-alignment of sequences. It takes a multi Fasta containing a list of sequences and returns an alignment.

Workflow w_u shows an ideal view where users have specified only the indispensable information to describe what has to be achieved. However, this view is not directly executable because the workflow uses services whose inputs and outputs do not immediately match. It is necessary to insert shims services to address data mismatches as generally done in the existing platforms.

Figure 6 shows an executable workflow (w_x) where shims have been inserted because of the following mismatches. First mismatch (tabular2multiFasta): the concrete lists of sequences the workflow will consume are in tabular form with additional columns, they must be adapted to feed the input of *clustalW*. Second mismatch (clustalAln2MSF): the output of task *clustalW* must be adapted to

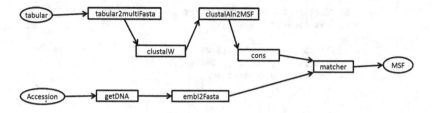

Fig. 6. Executable workflow (w_x).

match the input of task *cons*. Last mismatch (embl2Fasta): task *getDNA* provides an output that must be fed into the input of task *matcher*, the output of *getDNA* contains more information and is more specific. In existing approaches, to obtain an executable workflow, users must find and insert format converters between domain tasks for which the formats are different. Data are seen through their formats and they are not decomposable. There is no separation between data, formats and services and no separation between domain services and shim services.

Figure 7 shows the abstract workflow generated by our system (w_a). Ellipses are data types, boxes are domain tasks, and diamonds are generated converters used as shims. Our system sees service inputs and outputs through their composite abstract types. Each service is represented with its input and output types as follows:

– *Alignment* represents the output type of the service matcher, the input type of the service cons and the output type of the service clustalW. It also represents the output type of the workflow.
– *CBioSeq* represents the two input types of the service matcher and the output type of the service cons.
– *TFasta = seqs[id[string]   seq[Bioseq]]+* represents the input type of the service clustalW.
– *Accession* represents the input type of the service getDNA, and also an input type of the workflow.
– *CDNAFeatSeq= seq[CDNASeq   Features]* represents the output type of getDNA.
– *TTab=seqs[id[string]   C2[string]   C3[string]   C4[string]   seq[Bioseq]]+* represents an input type of the workflow.

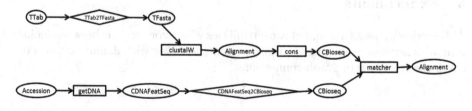

Fig. 7. Abstract workflow (w_a) in our system.

```
59  declare function my:convert($v0 as node()*)
60    as node()*
61  {
62  let $v1:=my:content($v0)
63  let $v2:=my:select('CDNASeq',$v1)
64  let $v3:=my:content($v2)
65  return $v3
66  };
```

Fig. 8. Example of source code of the shim converting data matching *CDNAFeatSeq* to data matching *CBioseq*. It is simplified and represented using XQuery. Functions *my:content* and *my:select* are utility functions, their source code is not shown here.

Our system separates data types, formats and services. It detects and resolves mismatches letting users focus on domain tasks. The generated shims fix the data mismatches presented above. The first generated shim, *TTab2TFasta*, corresponds to the resolution of the tabular2multiFasta mismatch, by transforming each element (row) of the list (table). Transforming each element involves selecting and concatenating sub-elements (fields). The second shim, *CDNAFeatSeq2CBioseq*, fixes the embl2Fasta mismatch by recognizing a DNA sequence as a biological sequence after ignoring additional information. The clustalAln2MSF mismatch is fixed by reflexivity because clustalAln and MSF correspond to the same abstract type. By ignoring derivable and optional information our system recognises them as equal. Figure 8 shows a generated XQuery source code of shim *CDNAFeatSeq2CBioseq*. It is similar to the theoretical example provided at Fig. 2.

From our abstract workflow (w_a), it is possible to obtain the workflow at user level (w_u) by hiding intermediate types and generated shims. It is also possible to produce the executable workflow (w_x) by inserting converters between concrete formats and XML (e.g., XML from/to each of tabular, multi Fasta, embl and Fasta).

Note that although the users do not have to interfere to generate the converters, they can still check them because all steps of the generation are traceable. In our system, formats are concrete serialization of data, they are not bound to particular data or services. Their use is flexible, for example changing input/output formats do not affect service compatibility or service implementation.

5 Experiments

This section presents a graph of convertibilities where connections between bioinformatics services are detected. It also presents a survey with domain experts to check the relevance of graph connections.

5.1   Convertibility Between Bioinformatics Services

We selected 30 services from platforms EMBOSS [3], EBI (European Bioinformatics Institute) [27], BioMoby [28] and services adopting the BioXSD format. Other variations and similar categories are provided in platforms but, as mentioned above, their input and output types do not change in general, and they would add little to our experiment. Tables 2 and 3 show examples of services and types.

From our selection of services, using our matching algorithm, we automatically generated a graph of the connections between services. Each connection is a convertibility, proved by our rule system, from an output type of a service to an input type of another service. Figure 9 shows an excerpt from the obtained graph. The complete graph and service list are available online[8]. The graph shows services, input/output types and conversions between types. Services are represented by rectangles, input/output types by ellipses and conversions by diamonds. Services are associated with their inputs and outputs respectively by incoming and outgoing arrows. Similarly, each conversion is associated with a source type and target type, it materializes an automatically detected conversion between two types. Connectivity of services in the graph materializes processing chains where output data are transformed according to the need of the service inputs. Conversions from external formats to our representation are not shown in the graph. Our algorithm finds

Table 2. Examples of services

Service	Source	Inputs	Outputs	Task
Blast	BioXSD	Biosequence database URI	Biosequence	Search
Blastp	EMBL-EBI	Fasta sequence (typed) database URI	BlastResult	Search
ClustalWFastaCollection	BioMoby	Fasta files	MSF	Alignment
ClustalW	BioXSD	Biosequence (≥ 2)	Alignment	Alignment
Maskfeat	EMBOSS	EMBL sequence	Fasta sequence	Handling

Table 3. Examples of formats and types

Type	Type in our representation	Represents
Fasta sequence	ComplexProteinSequence | ComplexBiosequence	(Typed) sequences
BlastResult, Fasta files	ListOfComplexBiosequences | ListOfBiosequences	A list of sequences
EMBL sequence	AnnotatedSequence	An annotated sequence
BioXSD biosequence	Biosequence | ComplexBiosequence	One sequence
BioXSD Alignment, MSF	SequenceAlignment	An alignement of sequences
Database URI	DatabaseReference	A reference to a database

---

[8] http://www.irisa.fr/LIS/Members/moba/graph/view.

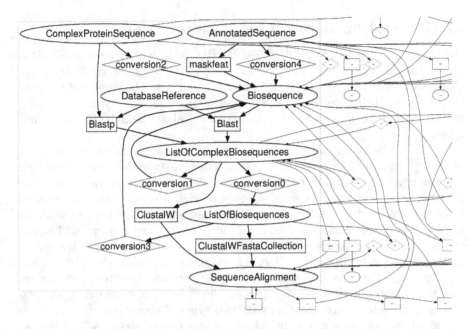

Fig. 9. Excerpt of the graph of links between services

direct links when the services use the same representation (the same type) to define the same data. This is the case, for example, with the link between services *Blast* and *ClustalW*. Indirect links correspond to a *conversion$_i$*. In the following, each *conversion$_i$(A, B)* specifies a function to transform data of type *A* to data of type *B*. With *conversion0(ListOfComplexBiosequences, ListOfBiosequence)*, a list of simple sequences is derived from a list of complex sequences. The function converts each element of the list and produces a new list of the converted elements. The elements of the list are converted using Rule (LEFTSELECTION) (or Rule (RIGHTSELECTION)), and the new list is produced by Rule (MAP). *Conversion1(ListOfComplexBiosequences, Biosequence)* combines rules (CHOICE) and (LEFTSELECTION) (or (RIGHTSELECTION)) to go from a list of complex protein sequences to each simple biological sequence of the list. A simple biological sequence being a component of a complex biological sequence. *Conversion2(ComplexProteinSequence, Biosequence)* shows the generalization and specialization of types. Our algorithm detects that a sequence of proteins is also a biological sequence. The conversion uses (LEFTPOSTCHOICE) (or (RIGHTPOSTCHOICE)) to obtain a complex biological sequence from the complex protein sequence and uses Rule (LEFTSELECTION) (or (RIGHTSELECTION)) to obtain a biological sequence from the complex biological sequence. Our algorithm also individually considers the elements of a list. With *conversion3(ListOfBiosequences, Biosequence)* each element of a list of sequences can be selected, it is done by rule (CHOICE). With *conversion4(AnnotatedSequence, Biosequence)*, a simple sequence is derived from an annotated sequence. A composite type is decomposed and some components are selected to feed services. This conversion corresponds to rules (LEFTSELECTION) and (RIGHTSELECTION). In above conversions, rules (TAGCHANGE), (TAGREMOVAL) and (PRIMITIVE) are used to convert between primitives and tags.

The complete graph contains 264 links between services out of the 900 possible links between 30 services (30×30). Our program therefore finds numerous links, but remains specific enough to be useful. Among those links, 88 are direct links, i.e., do not imply any conversion. Our convertibility relation therefore enables a three-fold increase of the number of links between services. The 30 services use 26 different types, among which 10 types are in fact ad-hoc and not decomposable (e.g., pictures, reports). Our program has identified 27 possible conversions between the 16 composite types, out of the 256 possible ones (16×16). This again shows that our approach is both productive and specific.

We presented the graph of the experiment to developers and users of the GenOuest bioinformatics platform. They pointed out that "it is remarkable that the central role of sequence alignment is so visible in the graph". They stated that "the produced graph has a pedagogical interest". Indeed, in a typical course handout[9], a graph produced by hand related to a library of bioinformatics services contains similar services and connections as our graph. However, being more complete in the modelling of input and output data, our approach offers more flexibility on input and output types. Thus our algorithm provides more explicit connections and differentiation between categories of services. It also reveals possible conversions between input and output data, that creates new connections. Moreover, our graph is machine processable, it shows a proof of the connections created between services and can quickly take into account new changes on data types and services. With a growing set of currently over 1500 available tools, it is unlikely that people can produce the graph by hand. Our main objective is to guide biologists when composing workflows. One perspective is to take into account other aspects of services. When constructing a workflow, input and output types play a central role in the selection of services by setting constraints on applicable services, but are not the only criteria for selection biologist. It is also necessary to represent the services by their functional and non-functional properties (e.g., bioinformatics task performed, quality of results, provenance, efficiency, popularity). The GenOuest developers nevertheless mentioned that a user with domain knowledge would already find useful support in the produced graph to select services for a workflow among the possibilities given by the graph. Thus, they validated that the graph generated by our approach detects relevant information and produces, in a systematic way, knowledge usually acquired by experience. The costly step of the approach is the production of abstract types, currently done by hand. They highlighted some interesting perspectives. Our data abstractions could be enriched from ontologies, especially EDAM, which will significantly facilitate the type abstraction step and allow integrating others facets. In addition, data in Genomics (e.g., phylogenetic trees) and in others domains (e.g., metabolism) have to be added to the experiment.

---

[9] Presentation of services of Wisconsin Package- Olivier Collin - CNRS Roscoff - Formation Génopole Ouest - november 2002.

5.2   Survey with Domain Experts

We evaluated the graph produced at Sect. 5.1 by asking domain experts to judge about the relevance of generated connections between services. For two services s_1 and s_2 in the graph, if the type of an output port of s_1 is convertible to the type of an input port of s_2, we suggested to experts to connect the input of s_2 to the output of s_1. We prepared 25 questions, each question being a suggestion to connect two services. Suggestions involved 18 services. The 25 questions concerned all connections on ports of 3 services (output port of blastp, input port of blast and input port of cons). For each question, it was possible to give one among 6 answers: "Yes, I knew", "Yes, I discover", "It may be feasible", "It is not feasible", "No, I disagree", and "I do not know". For questions we provided the description of services and a brief explanation of convertibility that involved the suggestion. To the 25 questions, we added general questions about user status, background knowledge, the services they know by their names, and their opinion about the relevance of suggestions. The survey was submitted to experts of bioinformatics data, services and workflows via a mailing list.

We collected responses from 6 participants. Among the participants, one did no provide any answer on suggestions, and therefore has been removed from the results on suggestions. One did not finish the survey, thus for him we just consider answered questions. All participants work in bioinformatics. Four of them are computer scientists, four are users of services and workflows, four are developers of workflows, five are developers of services, one is a developer of Workflow Management Systems (WfMS) and one is a researcher. Each participant knew on average 7 services among the 18 services proposed as suggestions. 3 of the services were known to all participants, 7 were known to no participants. The number of known services may look surprisingly low for domain experts. It can be explained by the fact that participants are actually expert in one among many bioinformatics platforms. The same task is often implemented by different services (with different names) across different platforms.

Figures 10 and 11 show the results of the survey. The table of Fig. 10 lists the responses for each participant and for each suggestion. The left column of the table contains suggestions established between services, for example 'blastp > *blast*' meant that the system suggested that service blast could consume the output of the service blastp; '*clustalW* < cons': the system suggested that clustalW could produce the input of the service cons. The second column of the table tells if connected services come from different platforms. Columns SL and SR contain the number of participants who know the tool of either the left part (SL) or the right part (SR). From column four to column eight, are the responses of the participants. The responses are represented by (++) for "Yes, I knew", (+) for "Yes, I discover", (o) for "It may be feasible", (-) for "It is not feasible", (--) for "No, I disagree" and (?) for "I do not know". Missing responses are represented by crosses. Figure 11 aggregates results. It provides the percentage by response type for all responses.

Suggested links	Cross-platforms	SL	SR	1st pers	2nd pers	3th pers	4th pers	5th pers
blastp > blast	no	5	6	--	-	++	+	o
blastp > blastp	no	5	5	o	-	++	+	o
blastp > pepcoil	yes	5	0	o	-	+	+	+
blastp > diffseq	yes	5	2	o	-	+	+	+
blastp > emma	yes	5	0	o	-	+	+	+
blastp > clustalW	yes	5	5	o	-	++	+	++
blastp > backtranseq	yes	5	0	+	-	+	+	++
blastp > clustalWFastaCollection	yes	5	2	+	-	+	+	++
blastp < blast	no	5	6	o	-	++	+	o
cons < blast	yes	0	6	+	+	+	o	o
blast < blast	no	6	6	++	-	++	++	o
merger < blast	yes	0	6	x	+	+	-	?
transeq < blast	yes	1	6	x	--	+	++	++
blastn < blast	no	6	6	x	-	++	++	o
blastx < blast	no	6	6	x	-	++	++	o
extractFeat < blast	yes	0	6	x	?	+	+	o
extractAlign < blast	yes	1	6	x	+	+	+	o
backtranseq < blast	yes	0	6	x	+	+	+	++
gassst < cons	yes	5	0	x	o	++	+	--
matcher < cons	no	3	0	x	+	o	+	o
maxAlign < cons	yes	0	0	x	o	+	+	o
emma < cons	no	0	0	x	o	+	+	o
merger < cons	no	0	0	x	+	+	+	o
clustalW < cons	yes	5	0	x	+	++	+	++
clustalWFastaCollection < cons	yes	2	0	x	+	++	+	++

Fig. 10. Responses of 5 participants to the list of link suggestions between services. 'Cross-platform' means services that are from different platforms. 'SL'/'SR' respectively count the number of participants who already knew the left/right service. Values in last the 5 columns are responses of participants for each suggestion. The responses are represented by (++) for "Yes, I knew", (+) for "Yes, I discover", (o) for "It may be feasible", (-) for "It is not feasible", (--) for "No, I disagree" and (?) for "I do not know". Crosses represent missing responses.

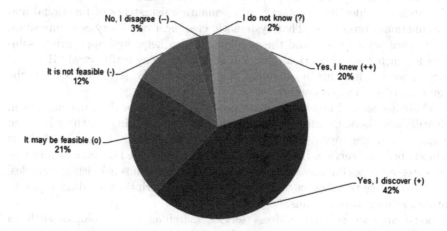

Fig. 11. Pie chat of all participant responses to all suggestions, per response type.

The results show that participants found that 83 % of suggested connections are acceptable. They already knew 20 % and discovered 42 % of suggested connections. They judged that 21 % of suggested connections may be feasible. However, they found that 12 % of suggested connections as not really feasible. They are in total disagreement with 3 % of suggested connections. The participants have no opinion for 2 % of suggested connections.

Despite knowing only a few (at most 15 %) of the suggested services, participants still recognized most suggestions (83 %) as acceptable. That paradoxical observation deserves an explanation. Many services are unknown to participants by their name, because belonging to another platform than the one they are used to, but their functions are known to them. For example, 'emma' has the same function as 'ClustalW', but only the latter is known by the participants. Thanks to the short description that we provided in the survey, and to online information, participants were able to recognize the function of unknown services, and to evaluate them accordingly. For example, the table of Fig. 10 shows that 'emma' is evaluated very similarly to 'ClustalW'. This also explains the high proportion of "Yes, I discover" responses (42 %), which is much higher than we expected. It implies two positive results. Firstly, our suggestions cover many relevant service connections that would not be considered by domain experts in the construction of a workflow by lack of knowledge about services. Secondly, most of our suggestions are recognized as relevant by domain experts.

6   Related Work

The development of automatic solutions for service composition is a response to the time-consuming and error prone methods currently used in some platforms to manage service selection and service mediation during composition of services. Mediating incompatible services requires identifying categories of mismatches. The work of Li et al. [11] provides a multi-dimensional classification of mismatches. It identifies syntactic and semantic mismatches of functional and non-functional properties. That systematic classification of service composition mismatches helps understand the problem. It also helps find appropriate solutions for each case (Sirin et al [29], Lin et al [30], Kongdenfha et al [31]). Our work concerns signature mismatches that occur on the structure and on the semantics of service parameters.

There are several approaches that provide solutions for data mismatches in scientific workflows. Besides approaches that address verifying matching between services, there are approaches that resolve data mismatches by means of shims to insert between services in workflows. They can be divided in two categories: A first category relies on semantic annotations to search shims in existing libraries. A second category takes into account syntactic descriptions of data types to automatically generate shims.

Some approaches that address service matching use ontologies such as EDAM [32] and myGrid ontology [33]. They provide methods to access semantic compatibility of workflow components. However, they do not generally guarantee compatibility at syntactic level. For example, Lebreton et al. [12] propose to verify the semantic compatibility of service parameters for web service composition. They address input and output matching but they do not provide solutions to resolve data mismatches that can occur on the data structure. Another approach, by Stroulia et al. [34], uses the structure of data types, messages, operations and textual descriptions to assess similarity between WSDL (Web Service Description Language) specifications. However, it is not designed to adapt data between services.

Compared to previous works, approaches that find shims address resolving data mismatches in workflows. They generally rely on data types, formats and service descriptions. For example, Velasco-Elizondo et al. [13] use data format descriptions to automatically identify relevant shims. In the same manner, the approach of Hull et al [14] relies on the description of shims and data types to retrieve shims transforming data between services. These approaches, besides seeing data as not decomposable, also expect the shims to be provided by third-parties.

In contrast to approaches that search shims, approaches that generate shims automatically provide data transformers to insert between services. They are characterized by the data complexity they support and the transformations between data they offer. Browers et al [15] support structural data transformations based on ontological information. Their approach links a structural type to a corresponding semantic type. The semantic type is defined with concepts of ontologies. However, the transformations that they offer rely on contextual paths. They do not allow transforming XML documents whose elements are not already attached to domain concepts. They also, do not seem to take into account recursion, which is required for some complex data. In Conveyor, Linke et al. [35] propose generic object-oriented type system to manage nodes, including data types, in workflows. They use interfaces, abstract classes and inheritances to express relations between types but do not mention automated methods ensuring composition and decomposition of complex data types.

Kashlev et al [16] use rules to automatically insert shims as coercions into executable workflows. Their approach and the approach of Dibernado et al [17] are close to our work. The first one because it uses rules, and the second one because it is applied to compose bioinformatics services. We differ from the approach of Kashlev et al. [16] on some points. We envision to mediate data at workflow construction time, not at execution time. We allow more complex data representations, for example type constructors union, tag and list. Thus, we offer additional transformations between input and output data. For example, the approach of Kashlev et al. [16] requires different labels for elements of a tuple while our abstraction authorizes to define a tuple where elements use the same label. This allows us to meet some bioinformatics data representations such as representing lists of biological sequences, and enables parallel transformation on lists. We differ from the approach of Dibernado et al. [17] because we allow decomposing as well as composing data. Dibernado et al. [17] provide the data transformation during workflow composition. They allow generating shims according to connected services. However, they decompose data but do not allow to compose them.

Compared to presented approaches, our approach offers richer type abstractions and more complex transformations between data. It relies on rules that allow to systematically reason on input and output types for managing mismatches during workflow composition. In addition, our approach does not prevent to use existing shims, it provides systematic and automatic mechanisms to re-use some of them (e.g., converters for upper case, lower case or substitution). Unlike many approaches, our approach does not only provide the solution in XML, it also provides mechanisms to link existing textual representations. Furthermore, our approach may easily benefit from mechanisms of matching based on ontologies.

Besides the automatic approaches, platforms such as Galaxy [4] use shims libraries to manage links between services. Matching between inputs and outputs is based on data formats. Services, in their implementation, take into account several formats. When a format is not provided, a format converter defined by hand is used. These techniques make a strong dependency between services and formats, which mixes domain tasks and tasks for the data adaptation. In addition, as discussed above, textual formats do not promote automation. Separating data types and formats is the subject of much work, for example Kalas et al. [9]. To facilitate interoperability of tools, common XML-based formats are proposed to represent bioinformatics data. At present, few implementations use these technologies. A generalization of their use would strengthen our approach, as it would facilitate the specification of abstract data types. Possibilities offered by XML technologies to represent complex data and relationships associated to domain ontologies may be used to provide a pivot language for conversion between heterogeneous data formats.

7   Perspectives

In the future, we plan to use our convertibility approach to propose an guided approach for composing real-world workflows. Our approach is appropriate for users of platforms such as Taverna [1] and Galaxy [4] that need to benefit more from automation. It can also be used to manage creation, re-use and conversion of parameters in workflow engines such as Bpipe [36], Snakemake [37] that use low-level programming. For our approach to be more beneficial, we will cope with data consistency and deal with non-determinism. We will integrate data consistency checking compared to types. Errors due to the non-compliance between data and constraints are common in automated platforms. For example, little changes in textual formats used to represent input data may cause a program to abort. An advantage of our approach is that it offers means to verify the compliance of data compared to their data type. Enriching abstractions with ontologies will allow to reduce non-determinism. It will be necessary, however, to integrate users in the process to manage multiple choices.

8   Conclusion

Data mismatches between services make it difficult to create scientific workflows. Manually defined shims proposed to fix data mismatches are time consuming and error prone. Existing approaches that automatically insert shim services in workflows are limited in the transformations they provide.

In this paper, we presented an approach that systematically detects convertibility from output types to input types. We have defined convertibility rules that exploit (de)composition as well as specialization and generalization of types. The rules also automatically generate converters between input and output XML data that can be used as shims. An experiment on bioinformatics services, as well as a survey with domain experts, showed that the detected convertibilities and produced converters are relevant from a biological point of view. Furthermore, the automatically produced graph of potentially compatible services exhibited a connectivity higher than with the ad'hoc approaches.

Acknowledgments. We thank Olivier Collin, Yvan Le Bras, Olivier Dameron, Francois Moreews and Olivier Sallou for their expertise in bioinformatics services and workflows, as well as for enriching discussions.

References

1. Oinn, T., Greenwood, M., Addis, M., Ferris, J., Glover, K., Goble, C., Hull, D., Marvin, D., Li, P., Lord, P.: Taverna: lessons in creating a workflow environment for the life sciences. Concurrency Comput. Pract. Experience **18**(10), 1067–1100 (2006)
2. Gundersen, S., Kalas, M., Abul, O., Frigessi, A., Hovig, E., Sandve, G.K.: Identifying elemental genomic track types and representing them uniformly. BMC Bioinform. **12**, 494 (2011)
3. Rice, P., Longden, I., Bleasby, A.: Emboss: the European molecular biology open software suite. Trends Genet. **16**(6), 276–277 (2000)
4. Goecks, J., Nekrutenko, A., Taylor, J., Team, T.G.: Galaxy: a comprehensive approach for supporting accessible, reproducible, and transparent computational research in the life sciences. Genome Biol. **11**(8), R86 (2010)
5. Ménager, H., Gopalan, V., Néron, B., Larroudé, S., Maupetit, J., Saladin, A., Tufféry, P., Huyen, Y., Caudron, B.: Bioinformatics applications discovery and composition with the mobyle suite and mobyleNet. In: Lacroix, Z., Vidal, M.E. (eds.) RED 2010. LNCS, vol. 6799, pp. 11–22. Springer, Heidelberg (2012)
6. Wassink, I.H.C., van der Vet, P.E., Wolstencroft, K., Neerincx, P.B.T., Roos, M., Rauwerda, H., Breit, T.M.: Analysing scientific workflows: why workflows not only connect web services. In: SERVICES, pp. 314–321 (2009)
7. Seibel, P.N., Krüger, J., Hartmeier, S., Schwarzer, K., Löwenthal, K., Mersch, H., Dandekar, T., Giegerich, R.: XML schemas for common bioinformatic data types and their application in workflow systems. BMC Bioinform. **7**, 490 (2006)
8. Han, M.V., Zmasek, C.M.: phyloXML: XML for evolutionary biology and comparative genomics. BMC Bioinform. **10**, 356 (2009)

9. Kalas, M., Puntervoll, P., Joseph, A., Bartaseviciute, E., Töpfer, A., Venkatara-man, P., Pettifer, S., Bryne, J.C., Ison, J.C., Blanchet, C., Rapacki, K., Jonassen, I.: Bioxsd: the common data-exchange format for everyday bioinformatics web services. Bioinformatics **26**(18), i540–i546 (2010)
10. Embley, D.W., Xu, L., Ding, Y.: Automatic direct and indirect schema mapping: experiences and lessons learned. SIGMOD Rec. **33**(4), 14–19 (2004)
11. Li, X., Fan, Y., Jiang, F.: A classification of service composition mismatches to support service mediation. In: GCC, pp. 315–321 (2007)
12. Lebreton, N., Blanchet, C., Claro, D.B., Chabalier, J., Burgun, A., Dameron, O.: Verification of parameters semantic compatibility for semi-automatic web service composition: a generic case study. In: Taniar, D., Pardede, E., Nguyen, H.-Q., Rahayu, J.W., Khalil, I. (eds.) International Conference on Information Integration and Web Based Applications and Services, pp. 845–848. ACM (2010)
13. Elizondo, P.V., Dwivedi, V., Garlan, D., Schmerl, B.R., Fernandes, J.M.: Resolving data mismatches in end-user compositions. In: IS-EUD, pp. 120–136 (2013)
14. Hull, D., Stevens, R., Lord, P., Wroe, C., Goble, C.: Treating "shimantic web" syndrome with ontologies (2004)
15. Bowers, S., Ludäscher, B.: An ontology-driven framework for data transformation in scientific workflows. In: Rahm, E. (ed.) DILS 2004. LNCS (LNBI), vol. 2994, pp. 1–16. Springer, Heidelberg (2004)
16. Kashlev, A., Lu, S., Chebotko, A.: Coercion approach to the shimming problem in scientific workflows. In: 2013 IEEE International Conference on Services Computing, Santa Clara, CA, USA, 28 June–3 July 2013, pp. 416–423 (2013)
17. DiBernardo, M., Pottinger, R., Wilkinson, M.: Semi-automatic web service composition for the life sciences using the biomoby semantic web framework. J. Biomed. Inform. **41**(5), 837–847 (2008)
18. Ba, M., Ferré, S., Ducassé, M.: Generating data converters to help compose services in bioinformatics workflows. In: Decker, H., Lhotská, L., Link, S., Spies, M., Wagner, R.R. (eds.) DEXA 2014, Part I. LNCS, vol. 8644, pp. 284–298. Springer, Heidelberg (2014)
19. Missier, P., Wolstencroft, K., Tanoh, F., Li, P., Bechhofer, S., Belhajjame, K., Pettifer, S., Goble, C.A.: Functional units: abstractions for web service annotations. In: SERVICES, pp. 306–313. IEEE Computer Society (2010)
20. Hosoya, H., Vouillon, J., Pierce, B.C.: Regular expression types for XML. In: ICFP, pp. 11–22 (2000)
21. Chen, Z., Wu, J., Deng, S., Li, Y., Wu, Z.: Describing and verifying web service using type theory. In: Proceedings of the 10th International Conference on CSCW in Design, CSCWD 2006, 3–5 May 2006, Southeast University, Nanjing, China, pp. 746–750 (2006)
22. Bates, J.L., Constable, R.L.: Proofs as programs. ACM Trans. Program. Lang. Syst. **7**(1), 113–136 (1985)
23. Moreews, F., Lavenier, D.: Seamless coarse grained parallelism integration in intensive bioinformatics workflows. In: 20th European MPI Users's Group Meeting, EuroMPI 2013, Madrid, Spain, 15–18 September 2013, pp. 277–282 (2013)
24. Westbrook, J.D., Ito, N., Nakamura, H., Henrick, K., Berman, H.M.: PDBML: the representation of archival macromolecular structure data in XML. Bioinformatics **21**(7), 988–992 (2005)
25. Dowell, R.D., Jokerst, R.M., Day, A., Eddy, S.R., Stein, L.: The distributed annotation system. BMC Bioinform. **2**, 7 (2001)
26. Consortium, U., et al.: The universal protein resource (uniprot) in 2010. Nucleic Acids Res. **38**, 142–148 (2010). Database-Issue

27. McWilliam, H., Valentin, F., Goujon, M., Li, W., Narayanasamy, M., Martin, J., Miyar, T., Lopez, R.: Web services at the European bioinformatics institute-2009. Nucleic Acids Res. **37**, 6–10 (2009). Web-Server-Issue
28. Wilkinson, M.D., Links, M.: Biomoby: an open source biological web services proposal. Briefings Bioinform. **3**(4), 331–341 (2002)
29. Sirin, E., Hendler, J., Parsia, B.: Semi-automatic composition of web services using semantic descriptions. In: Web Services: Modeling, Architecture And Infrastructure Workshop in ICEIS, vol. 2003. Citeseer (2003)
30. Lin, C., Lu, S., Fei, X., Pai, D., Hua, J.: A task abstraction and mapping approach to the shimming problem in scientific workflows. In: 2009 IEEE International Conference on Services Computing (SCC 2009), Bangalore, India, 21–25 September 2009, pp. 284–291 (2009)
31. Kongdenfha, W., Nezhad, H.R.M., Benatallah, B., Casati, F., Saint-Paul, R.: Mismatch patterns and adaptation aspects: a foundation for rapid development of web service adapters. IEEE T. Serv. Comput. **2**(2), 94–107 (2009)
32. Ison, J.C., Kalas, M., Jonassen, I., Bolser, D.M., Uludag, M., McWilliam, H., Malone, J., Lopez, R., Pettifer, S., Rice, P.M.: EDAM: an ontology of bioinformatics operations, types of data and identifiers, topics and formats. Bioinformatics **29**(10), 1325–1332 (2013)
33. Wolstencroft, K., Alper, P., Hull, D., Wroe, C., Lord, P.W., Stevens, R.D., Goble, C.A.: The myGrid ontology,: bioinformatics service discovery. Int. J. Bioinform. Res. Appl. **3**(3), 303–325 (2007)
34. Stroulia, E., Wang, Y.: Structural and semantic matching for assessing web-service similarity. Int. J. Coop. Inf. Syst. **14**(4), 407–438 (2005)
35. Linke, B., Giegerich, R., Goesmann, A.: Conveyor: a workflow engine for bioinformatic analyses. Bioinformatics **27**(7), 903–911 (2011)
36. Sadedin, S.P., Pope, B., Oshlack, A.: Bpipe: a tool for running and managing bioinformatics pipelines. Bioinformatics **28**(11), 1525–1526 (2012)
37. Köster, J., Rahmann, S.: Snakemake:a scalable bioinformatics workflow engine. Bioinformatics **28**(19), 2520–2522 (2012)

A Framework for Sampling-Based XML Data Pricing

Ruiming Tang[1], Antoine Amarilli[2], Pierre Senellart[1,2]([✉]),
and Stéphane Bressan[1]

[1] National University of Singapore, Singapore, Singapore
`tangruiming1987@gmail.com, steph@nus.edu.sg`
[2] Institut Mines–Télécom, Télécom ParisTech, CNRS LTCI, Paris, France
`{antoine.amarilli,pierre.senellart}@telecom-paristech.fr`

Abstract. While price and data quality should define the major trade-off for consumers in data markets, prices are usually prescribed by vendors and data quality is not negotiable. In this paper we study a model where data quality can be traded for a discount. We focus on the case of XML documents and consider completeness as the quality dimension.

In our setting, the data provider offers an XML document, and sets both the price of the document and a weight to each node of the document, depending on its potential worth. The data consumer proposes a price. If the proposed price is lower than that of the entire document, then the data consumer receives a sample, i.e., a random rooted subtree of the document whose selection depends on the discounted price and the weight of nodes. By requesting several samples, the data consumer can iteratively explore the data in the document.

We present a *pseudo-polynomial time* algorithm to select a rooted subtree with prescribed weight uniformly at random, but show that this problem is unfortunately intractable. Yet, we are able to identify several practical cases where our algorithm runs in polynomial time. The first case is uniform random sampling of a rooted subtree with prescribed size rather than weights; the second case restricts to binary weights.

As a more challenging scenario for the sampling problem, we also study the uniform sampling of a rooted subtree of prescribed weight and prescribed height. We adapt our pseudo-polynomial time algorithm to this setting and identify tractable cases.

1 Introduction

There are three kinds of actors in a data market: data consumers, data providers, and data market owners [14]. A data provider brings data to the market and sets prices on the data. A data consumer buys data from the market and pays for it. The owner is the broker between providers and consumers, who negotiates pricing schemes with data providers and manages transactions to trade data.

In most of the data pricing literature [4–6,9], data prices are prescribed and not negotiable, and give access to the best data quality that the provider can achieve. Yet, data quality is an important axis which should be used to price

© Springer-Verlag Berlin Heidelberg 2016
A. Hameurlain et al. (Eds.): TLDKS XXIV, LNCS 9510, pp. 116–138, 2016.
DOI: 10.1007/978-3-662-49214-7_4

documents in data markets. Wang et al. [15,19] define dimensions to assess data quality following four categories: intrinsic quality (believability, objectivity, accuracy, reputation), contextual quality (value-added, relevancy, timeliness, ease of operation, appropriate amount of data, completeness), representational quality (interpretability, ease of understanding, concise representation, consistent representation), and accessibility quality (accessibility, security).

In this paper, we focus on contextual quality and propose a data pricing scheme for *XML trees* such that *completeness* can be traded for discounted prices. This is in contrast to our previous work [18] where the *accuracy* of *relational data* is traded for discounted prices. Wang et al. [15,19] define completeness as "the extent to which data includes all the values, or has sufficient breadth and depth for the current task". We retain the first part of this definition as there is no current task defined in our setting. Formally, the data provider assigns, in addition to a price for the entire document, a *weight* to each node of the document, which is a function of the potential worth of this node: a higher weight is given to nodes that contain information that is more valuable to the data consumer. We define the completeness of a rooted subtree of the document as the total weight of its nodes, divided by the total weight of the document. A data consumer can then offer to buy an XML document for less than the provider's set price, but then can only obtain a rooted subtree of the original document, whose completeness depends on the discount granted.

A data consumer may want to pay less than the price of the entire document for various reasons: first, she may not be able to afford it due to limited budget but may be satisfied by a fragment of it; second, she may want to explore the document and investigate its content and structure before purchasing it fully. In this light, one may think of discounted samples of the complete documents as an inexpensive way for the user to discover which kind of content the document contains, so that she can make up her mind about whether she wishes to purchase the complete document.

The data market owner negotiates with the data provider a pricing function, allowing them to decide the price of a rooted subtree, given its completeness (i.e., the weight). The pricing function should satisfy a number of axioms: the price should be non-decreasing with the weight, be bounded by the price of the overall document, and be *arbitrage-free* when repeated requests are issued by the same data consumer (arbitrage here refers to the possibility to strategize the purchase of data). Hence, given a proposed price by a data consumer, the inverse of the pricing function decides the completeness of the sample that should be returned. To be fair to the data consumer, there should be an equal chance to explore every possible part of the XML document that is worth the proposed price. Based on this intuition, we sample a rooted subtree of the XML document of a certain weight, according to the proposed price, uniformly at random.

The data consumer may also issue repeated requests as she is interested in this XML document and wants to explore more information inside in an iterative manner. For each repeated request, a new rooted subtree is returned. A principle here is that the information (document nodes) already paid for should not be charged again. Thus, in this scenario, we sample a rooted subtree of the XML document of a certain weight uniformly at random, without counting the weight of the nodes already bought in previously issued requests.

The present article brings the following contributions:

- We propose to realize the trade-off between quality and discount in data markets. We propose a framework for pricing the completeness of XML data, based on uniform sampling of rooted subtrees of prescribed weight in weighted XML documents (Sect. 3).
- We show that the general uniform sampling problem in weighted XML trees is intractable. In this light, we propose two restrictions: sampling based on the number of nodes, and sampling when weights are binary (i.e., weights are 0 or 1) (Sect. 4).
- We propose a pseudo-polynomial time algorithm for the general uniform sampling problem on prescribed weight, with the proof of its correctness and complexity (Sect. 5).
- We show that the two restricted problem variants are tractable by showing that the pseudo-polynomial time algorithm for the general sampling problem runs in polynomial time for uniform sampling based on the size of a rooted subtree, or on 0/1-weights (Sect. 6).
- We extend our framework to the case of repeated sampling requests with the requirement that the data consumer is never charged twice for the same nodes. Again, we obtain tractability when the weight of a subtree is its size (Sect. 7).
- As a more challenge scenario, we study the uniform sampling problem on both prescribed weight and height. We devise a pseudo-polynomial time to solve this sampling problem and also identify tractable cases for which the pseudo-polynomial time sampling algorithm performs in polynomial-time (Sect. 8).

This article is the journal version of our previous work [17], extended with the pseudo-polynomial time algorithm for the general weighted sampling problem, and the problem of sampling for prescribed weight and height.

2   Related Work

Data Pricing. The basic structure of data markets and different pricing schemes were introduced in [14]. The notion of "query-based" pricing was introduced in [4,6] to define the price of a query as the price of the cheapest set of pre-defined views that can determine the query. It makes data pricing more flexible, and serves as the foundation of a practical data pricing system [5]. The price of aggregate queries has been studied in [9]. Different pricing schemes are investigated and multiple pricing functions are proposed to avoid several pre-defined arbitrage situations in [10]. However, none of the works above takes data quality into account, and those works do not allow the data consumer to propose a price less than that of the data provider, which is the approach that we study here.

The idea of trading off price for data quality has been explored in the context of privacy in [8], which proposes a theoretic framework to assign prices to noisy query answers. If a data consumer cannot afford the price of a query, she can choose to tolerate a higher standard deviation to lower the price. However, this work studies pricing on accuracy for linear relational queries, rather than pricing XML data based on completeness. In [18], we propose a relational data pricing framework in which data accuracy can be traded for discounted prices. By contrast, this paper studies pricing for XML data, and proposes a tradeoff based on data completeness rather than accuracy.

Subtree/Subgraph Sampling. The main technical result of this paper is the tractability of uniform subtree sampling under a certain requested size. This question is related to the general topic of subtree and subgraph sampling, but, to our knowledge, it has not yet been adequately addressed.

Subgraph sampling works such as [3,7,13,16] have proposed algorithms to sample small subgraphs from an original graph while attempting to preserve selected metrics and properties such as degree distribution, component distribution, average clustering coefficient and community structure. However, the distribution from which these random graphs are sampled is not known and cannot be guaranteed to be uniform.

Other works have studied the problem of uniform sampling [2,11]. However, [2] does not propose a way to fix the size of the samples. The authors of [11] propose a sampling algorithm to sample a connected sub-graph of size k under an approximately uniform distribution; note that this work provides no bound on the error relative to the uniform distribution.

Sampling approaches are used in [12,20] to estimate the selectivity of XML queries (containment join and twig queries, respectively). Nevertheless, the samples in [20] are specific to containment join queries, while those in [12] are representatives of the XML document for any twig queries. Neither of those works controls the distribution from which the subtrees are sampled.

In [1], Cohen and Kimelfeld show how to evaluate a deterministic tree automaton on a probabilistic XML document. This has applications to sampling possible worlds that satisfy a given constraint, e.g., expressed in monadic second-order logic and then translated into a tree automaton. Note that the translation of constraints to tree automata itself is not tractable in general; in this respect, our approach can be seen as a specialization of [1] to the simpler case of fixed-size, fixed-weight, or fixed-height tree sampling, and as an application of it to data pricing.

3   Pricing Function and Sampling Problem

This paper studies data pricing for tree-shaped documents. We start by formally defining the terminology that we use for such documents.

We consider trees that are unordered, directed, rooted, and weighted; we equivalently call them *XML documents*. Formally, a tree t consists of a set of nodes $V(t)$ (which are assumed to carry unique identifiers), a set of edges $E(t)$, and a function w mapping every node $n \in V(t)$ to a non-negative rational number $w(n)$ which is the *weight* of node n. We write $root(t)$ for the root node of t. Whenever two nodes n_1, $n_2 \in V(t)$ are such that $(n_1, n_2) \in E(t)$, we say that n_1 and n_2 are in a *parent-child relationship*, that is, n_1 is the parent of n_2 and n_2 is a child of n_1.

By $children(n)$, we represent the set of nodes that have parent n. A tree is said to be *binary* if each node of the tree has at most two children, otherwise it is *unranked*.

We now introduce the notion of *rooted subtree* of an XML document:

Definition 1. *(Subtree, rooted subtree) A tree t' is a subtree of a tree t if $V(t') \subseteq V(t)$ and $E(t') \subseteq E(t)$. A rooted subtree t' of a tree t is a subtree of t such that $root(t) = root(t')$. We name it r-subtree for short. The weight function for a subtree t' of a tree t is always assumed to be the restriction of the weight function for t on the nodes in t'.*

For technical reasons, we also sometimes talk of the *empty* subtree that contains no node.

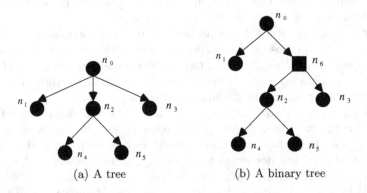

(a) A tree   (b) A binary tree

Fig. 1. Two example trees (usage of the square node will be introduced in Sect. 5.2)

Example 1. Figure 1 presents two example trees. The nodes $\{n_0, n_2, n_5\}$, along with the edges connecting them, form an r-subtree of the tree in Fig. 1(a). Likewise, the nodes $\{n_2, n_4, n_5\}$ and the appropriate edges form a subtree of that tree (but not an r-subtree). The tree of Fig. 1(b) is a binary tree (ignore the different shapes of the nodes for now).   □

We now present our notion of data quality, by defining the completeness of an r-subtree, based on the weight function of the original tree:

Definition 2. *(Weight of a tree) For a node $n \in V(t)$ of a tree t, we define inductively $weight(n) := w(n) + \sum_{(n,n') \in E(t)} weight(n')$. With slight abuse of notation, we note $weight(t) := weight(root(t))$ as the weight of t.*

Definition 3. *(Completeness of an r-subtree) Let t be a tree and t' be an r-subtree of t. The* completeness *of t' with respect to t is $c_t(t') := \frac{\text{weight}(t')}{\text{weight}(t)}$. It is obvious that $c_t(t') \in [0,1]$.*

We study a framework for data markets where the data consumer can buy an incomplete document from the data provider while paying a discounted price. The formal presentation of this framework consists of three parts:

1. An XML document t.
2. A pricing function φ_t for t whose input is the desired completeness for an r-subtree of the XML document, and whose value is the price of this r-subtree. Hence, given a proposed price pr_0 by a data consumer, the completeness of the returned r-subtree is decided by $\varphi_t^{-1}(pr_0)$.
3. An algorithm to sample an r-subtree of the XML document uniformly at random among those of a given completeness.

We study the question of the sampling algorithm more in detail in subsequent sections. For now, we focus on the pricing function, starting with a formal definition:

Definition 4. *(Pricing function) The* pricing function *for a tree t is a function $\varphi_t : [0,1] \to \mathbb{Q}^+$. Its input is the completeness of an r-subtree t' and it returns the price of t', as a non-negative rational.*

A healthy data market should impose some restrictions on φ_t, such as:

Non-decreasing. The more complete an r-subtree is, the more expensive it should be, i.e., $c_1 \geqslant c_2 \Rightarrow \varphi_t(c_1) \geqslant \varphi_t(c_2)$.

Arbitrage-free. Buying an r-subtree of completeness $c_1 + c_2$ should not be more expensive than buying two subtrees with respective completeness c_1 and c_2, i.e., $\varphi_t(c_1) + \varphi_t(c_2) \geqslant \varphi_t(c_1 + c_2)$. In other words, φ_t should be sub-additive. This property is useful when considering repeated requests, studied in Sect. 7.

Minimum and maximum bound. We should have $\varphi_t(0) = pr_{\min}$ and $\varphi_t(1) = pr_t$, where $pr_{\min}$ is the minimum cost that a data consumer has to pay using the data market and pr_t is the price of the whole tree t. Note that by the non-decreasing character of φ_t, $pr_t \geqslant pr_{\min} \geqslant 0$.

All these properties can be satisfied, for instance, by functions of the form $\varphi_t(c) := (pr_t - pr_{\min})c^p + pr_{\min}$ where $p \leqslant 1$; however, if $p > 1$, the arbitrage-free property is violated.

Given a proposed price pr_0 by a data consumer, $\varphi_t^{-1}(pr_0)$ is the set of possible corresponding completeness values. Note that φ_t^{-1} is a relation and may not be a function; φ_t^{-1} is a function if different completeness values correspond to different prices. Once a completeness value $c \in \varphi_t^{-1}(pr_0)$ is chosen, the weight of the returned r-subtree is fixed as $c \times \text{weight}(t)$.

Therefore, in the rest of the paper, we consider the problem of uniform sampling an r-subtree with prescribed weight (instead of with prescribed completeness). We now define the problem that should be solved by our sampling algorithm:

Definition 5. *(Sampling problem) The problem of* sampling an r-subtree, *given a tree t and a weight k, is to sample an r-subtree t′ of t, such that* weight(t′) = k, *uniformly at random, if one exists, or to fail if no such r-subtree exists.*

4   Tractability

Having defined our sampling problem, we now turn to the question of designing an algorithm to solve it, and of studying its complexity.

4.1   Intractability of the Sampling Problem

We start by showing that this problem is NP-hard in the general formulation that we gave.

Proposition 1. *Given a tree t and a weight k, it is* NP-*complete to decide whether there exists an r-subtree of weight k, and* NP-*hard to sample such an r-subtree uniformly at random.*

Proof. Deciding whether there exists an r-subtree of weight k is in NP, since, given an r-subtree, it takes polynomial time to check whether this r-subtree is of weight k by summing up the weights of all the nodes.

We now show that the problem is NP-hard, by describing a PTIME reduction from the NP-hard subset-sum problem. This is the problem of determining, given a set S of integers and a target value v (written in binary), whether there exists a subset $S′ \subseteq S$ which sums to v. Any set S can be encoded in polynomial time to a tree t such that $w(n) = 0$ except if n is a leaf, and the leaves correspond to the elements of S. Now, clearly there is an r-subtree of weight v in t iff there is a subset of S with sum v. This completes the reduction and shows that the problem of deciding the existence of an r-subtree of weight k is NP-complete.

Now there is a PTIME-reduction from the decision problem to the sampling problem, as an algorithm for sampling can be used to decide whether there exists an r-subtree of the desired weight (the algorithm returns one such) or if none exists (the algorithm fails). Therefore, the sampling problem is NP-hard.   □

Even though the general sampling problem is intractable, we devise in Sect. 5 a *pseudo-polynomial time* algorithm to solve it, which runs in polynomial time in the *value* of k (but is exponential in the size of k).

4.2   Tractable Cases

We now define restricted variants of the sampling problem where the weight function is assumed to be of a certain form. In Sect. 6, we show that sampling for these variants can be performed in PTIME.

Unweighted Sampling. In this setting, we take the weight function $w(n) = 1$ for all $n \in V(t)$. Hence, the weight of a tree t is actually the number of nodes in t, i.e., its size, which we write $\mathsf{size}(t)$.

In this case, the hardness result of Proposition 1 does not apply anymore. However, sampling an r-subtree with prescribed size uniformly at random is still not obvious to do, as the following example shows:

Example 2. Consider the problem of sampling an r-subtree t' of size 3 from the tree t in Fig. 1(a). We can enumerate all such r-subtrees: $\{n_0, n_1, n_2\}$, $\{n_0, n_1, n_3\}$, $\{n_0, n_2, n_3\}$, $\{n_0, n_2, n_4\}$ and $\{n_0, n_2, n_5\}$, and choose one of them at random with probability $\frac{1}{5}$. However, as the number of r-subtrees may be exponential in the size of the document in general, we cannot hope to perform this approach in PTIME.

Observe that it is not easy to build a random r-subtree node by node: it is clear that node n_0 must be included, but then observe that we cannot decide to include n_1, n_2, or n_3 uniformly at random. Indeed, if we do this, our distribution on the r-subtrees will be skewed, as n_1 (or n_3) occurs in $\frac{2}{5}$ of the outcomes whereas n_2 occurs in $\frac{4}{5}$ of them. Intuitively, this is because there are more ways to choose the next nodes when n_2 is added, than when n_1 or n_3 are added. □

0/1-weights Sampling. In this problem variant, we require that $w(n) \in \{0, 1\}$ for all $n \in V(t)$, i.e., the weight is a binary value. This variant generalizes the unweighted sampling case, but allows the data provider to give a weight of zero to some nodes that she is willing to give away for free. This can be useful, e.g., for nodes that are only structural and do not contain any useful information.

5 Algorithms for General Sampling Problem

In this section, we present a pseudo-polynomial algorithm for the general sampling problem, namely the problem of sampling an r-subtree of weight k from an XML document, uniformly at random.

We first describe the algorithm for the case of *binary* trees, in Sect. 5.1. Next, we adapt the algorithm in Sect. 5.2 to show how to apply it to arbitrary trees.

5.1 Sampling for Binary Trees

In this section, we provide an algorithm which proves the following theorem:

Theorem 1. *The sampling problem for binary trees can be solved in time $O(nk^2)$, where n is the number of nodes in the tree and k is the desired weight value.*

Our general algorithm to solve this problem is given as Algorithm 1. The algorithm has two phases, which we study separately in what follows. For simplicity, whenever we discuss binary trees in this section, we will add special

---

Algorithm 1. Algorithm for the sampling problem on binary trees

---

    Input: a binary tree t and an integer $k \geqslant 0$
    Result: an r-subtree t' of t of weight$(t') = k$ uniformly at random
    `// Phase 1: count the number of subtrees`
1 $D \leftarrow$ SubtreeCounting(t);
    `// Phase 2: sample a random subtree`
2 **if** $k \leqslant$ weight$(t) \land D_{\text{root}(t)}[k] \neq 0$ **then**
3 | **return** UniformSampling$(\text{root}(t), D, k)$;
4 **else**
5 | **fail** ;

NULL children to every node of the tree (except NULL nodes themselves), so that all nodes, including leaf nodes (but excluding NULL nodes), have exactly two children (which may be NULL). This will simplify the presentation of the algorithms. Of course weight(NULL) = 0.

First Phase: Subtree Counting (Algorithm 2). We start by computing a matrix D such that, for every node n_i of the input tree t and any value $0 \leqslant x \leqslant$ weight(t), $D_i[x]$ is the number of subtrees of weight x rooted at node n_i. We do so with Algorithm 2 which we now explain in detail.

There is only one subtree rooted at the special NULL node, namely the empty subtree, with weight 0, which provides the base case of the algorithm (line 3). Otherwise, we compute D_i for a node n_i from the values D_l and D_r of D for its children n_l and n_r (which may be NULL); those values have been computed before because nodes are considered bottom-up.

Intuitively, any r-subtree of weight $x > 0$ rooted at n_i is obtained by retaining n_i, and choosing two r-subtrees t_l and t_r, respectively rooted at n_l and n_r (the children of n_i), such that weight(t_l) + weight$(t_r) = x - \mathsf{w}(n_i)$ (which accounts for the weight of the additional node n_i). The number of such choices is computed by the *convolution* of D_l and D_r in line 7, defined as:

$$\text{For } 0 \leqslant p \leqslant \text{weight}(t), \quad (D_l * D_r)[p] := \sum_{m=0}^{p} D_l[m] \times D_r[p - m].$$

We explain Algorithm 2 by considering the cases of $\mathsf{w}(n_i) = 0$ (line 8 to line 11) and of $\mathsf{w}(n_i) \neq 0$ (line 13 to line 15), respectively.

If $\mathsf{w}(n_i) = 0$, the number of r-subtrees of weight $x > 0$ rooted at n_i is the number of pairs of r-subtrees t_l and t_r, respectively rooted at n_l and n_r (the children of n_i), such that weight(t_l) + weight$(t_r) = x$ (because node n_i does not contribute weight). That is to say, $D_i[x] = (D_l * D_r)[x]$ for $x > 0$ (line 11). By contrast, for $x = 0$, an r-subtree of weight 0 rooted at n_i can be obtained either in the same way, or by keeping the empty subtree. Therefore $D_i[0] = 1 + (D_l * D_r)[0]$ (line 9).

If $\mathsf{w}(n_i) \neq 0$, there is only one r-subtree of weight 0 at n_i, namely the empty tree. That is to say, $D_i[0] = 1$ as shown in line 13. The number of r-subtrees

---

Algorithm 2. SubtreeCounting(t)

Input: a binary tree t
Result: a matrix D such that $D_i[x]$ is the number of r-subtrees of weight x
rooted at n_i for all n_i and x

1 **for** $x \in [0, \mathsf{weight}(t)], n_i \in V(t) \sqcup \{\mathsf{NULL}\}$ **do**
2 $\quad$ $D_i[x] \leftarrow 0$;
3 $D_{\mathsf{NULL}}[0] \leftarrow 1$;
   $\quad$ // We browse all nodes in topological order, from leaves to the root
4 **foreach** *non*-NULL *node n_i accessed bottom-up* **do**
5 $\quad$ $n_l \leftarrow$ first child of n_i;
6 $\quad$ $n_r \leftarrow$ second child of n_i;
7 $\quad$ $T \leftarrow D_l * D_r$;
8 $\quad$ **if** $\mathsf{w}(n_i) = 0$ **then**
9 $\quad\quad$ $D_i[0] \leftarrow 1 + T[0]$;
10 $\quad\quad$ **for** $x \in [1, \mathsf{weight}(n_i)]$ **do**
11 $\quad\quad\quad$ $D_i[x] \leftarrow T[x]$;
12 $\quad$ **else**
13 $\quad\quad$ $D_i[0] \leftarrow 1$;
14 $\quad\quad$ **for** $x \in [\mathsf{w}(n_i), \mathsf{weight}(n_i)]$ **do**
15 $\quad\quad\quad$ $D_i[x] \leftarrow T[x - \mathsf{w}(n_i)]$;
16 **return** D;

---

of weight $x \geqslant \mathsf{w}(n_i)$ rooted at n_i is the number of pairs of r-subtrees t_l and t_r rooted at n_l and n_r (the children of n_i) respectively such that $\mathsf{weight}(t_l) + \mathsf{weight}(t_r) = x - \mathsf{w}(n_i)$, which implies that $D_i[x] = (D_l * D_r)[x - \mathsf{w}(n_i)]$ (line 15). For $x \in [1, \mathsf{w}(n_i) - 1]$, it is impossible to get an r-subtree of weight x at n_i, so $D_i[x]$ remains at 0.

Example 3. Let t be the tree presented in Fig. 1(b) (again, ignore the different shapes of nodes for now). Assume $\mathsf{w}(n_0) = \mathsf{w}(n_1) = \mathsf{w}(n_2) = 0$, $\mathsf{w}(n_3) = \mathsf{w}(n_4) = 1$, $\mathsf{w}(n_5) = \mathsf{w}(n_6) = 2$. Starting from the leaf nodes, we compute $D_4 = D_3 = (1, 1)$ and $D_5 = (1, 0, 1)$, with $T = D_{\mathsf{NULL}} * D_{\mathsf{NULL}} = (1)$, by line 13 to line 15. We compute $D_1 = (2)$ with line 8 to line 11.

Now, when computing D_2, we first convolve D_4 and D_5 to get the numbers of pairs of r-subtrees of different weights at $\{n_4, n_5\}$, i.e., $D_4 * D_5 = (1, 1, 1, 1)$, so that $D_2 = (2, 1, 1, 1)$ (applying line 8 to line 11). When computing D_6, we first compute $D_2 * D_3 = (2, 3, 2, 2, 1)$, so that $D_6 = (1, 0, 2, 3, 2, 2, 1)$ (applying line 13 to line 15). Finally, $D_0 = (3, 0, 4, 5, 4, 4, 3)$. $\qquad\square$

We now state the correctness and running time of this algorithm.

Lemma 1. *Algorithm 2 terminates in $O(nk^2)$ time (where n is the number of nodes in the given tree, and k is the desired weight) and returns D such that, for every n_i and x, $D_i[x]$ is the number of r-subtrees of weight x rooted at node n_i.*

Proof. We first prove the running time. All arrays under consideration have size at most W (where $W = \mathsf{weight}(t)$), so computing the convolution of two such arrays is in time $O(W^2)$. The number of convolutions to compute overall is $O(n)$, because each array D_i occurs in exactly one convolution. The overall running time is thus $O(nW^2)$. The running time can be optimized because, in fact, the arrays in D do not need to have size W, but can be bounded to size k (subtrees larger than k are never relevant). Therefore the complexity is in fact $O(nk^2)$.

We now show correctness. We proceed by induction on the node n_i to prove the claim for every x. The base case is the NULL node, whose correctness is straightforward. To prove the induction step, let n_i be a node, and assume by induction that $D_l[x']$ is correct for every x' and every child n_l of n_i. We fix x and show that $D_i[x]$ is correct.

We distinguish two cases in the proof: $w(n_i) \neq 0$ and $w(n_i) = 0$.

We first prove the case where $w(n_i) \neq 0$. To select an r-subtree at n_i, (1) if $x = 0$, there is exactly one possibility (the empty subtree); (2) if $x \in [1, w(n_i))$, there is no such possibility; (3) if $x \geq w(n_i)$, the number of possibilities is the number of ways to select a pair of r-subtrees at the children of n_i so that their weights sum to $x - w(n_i)$. This is the role of line 13 to line 15.

Now, to enumerate the ways of choosing r-subtrees at children of n_i whose weight sum to $x - w(n_i)$, we can first decide the weight of the selected r-subtree for each child: the ways to assign such weights form a partition of the possible outcomes, so the number of outcomes is the sum, over all such assignments of r-subtree weights to children, of the number of outcomes for this assignment. For a fixed assignment, the subtrees rooted at each children are chosen separately, so the number of outcomes for a fixed assignment is the product of the number of outcomes for the given weight for each child, which by induction hypothesis is correctly reflected by the corresponding $D_l[x']$. Hence, for a given x, (1) when $x = 0$, $D_i[x] = 1$ (line 13); (2) when $x \in [1, w(n_i))$, $D_i[x] = 0$ (from how D_i is initialized); (3) when $x \geq w(n_i)$, $D_i[x]$ is $(D_l * D_r)[x - w(n_i)]$ by line 15, which sums, over all possible subtree weights assignments, the number of choices for this subtree weight assignment.

We next prove the case where $w(n_i) = 0$. To select an r-subtree at n_i, (1) if $x = 0$, the number of possibilities is the number of ways to select a pair of r-subtrees at the children of n_i so that their weights sum to 0 plus one (this extra possibility is the empty subtree); (2) if $x > 0$, the number of possibilities is the number of ways to select a pair of r-subtrees at the children of n_i so that their weights sum to x, because node n_i contributes no weight. Similar to the previous case, the number of ways to select a pair of r-subtrees at the children of n_i so that their weights sum to x is $(D_l * D_r)[x]$. Therefore, for a given x, (1) when $x = 0$, $D_i[x] = 1 + (D_l * D_r)[0]$ (line 9); (2) when $x \neq 0$, $D_i[x] = (D_l * D_r)[x]$ (line 11).

Hence, by induction, we have shown the desired claim.   □

Second Phase: Uniform Sampling (Algorithm 3). In the second phase of Algorithm 1, we sample an r-subtree from t in a recursive top-down manner, based on the matrix D computed by Algorithm 2. Our algorithm to perform this uniform sampling is Algorithm 3. The basic idea is that to sample an r-subtree

---

Algorithm 3. UniformSampling(n_i, D, x)

Input: a node n_i (or NULL), the precomputed D, and a weight value x
Result: an r-subtree of weight x at node n_i

1 **if** $x = 0 \wedge w(n_i) \neq 0$ **then**
2 $\quad \lfloor$ **return** $\emptyset$;

3 $p_0 \leftarrow$ rand($[0, 1]$); // p_0 is generated from $[0, 1]$ randomly
4 **if** $x = 0 \wedge w(n_i) = 0 \wedge p_0 \leqslant \frac{1}{D_i[0]}$ **then**
5 $\quad \lfloor$ **return** $\emptyset$;

6 $n_l \leftarrow$ first child of n_i;
7 $n_r \leftarrow$ second child of n_i;
8 **for** $0 \leqslant s_l, s_r \leqslant x$ s.t. $s_l + s_r = x - w(n_i)$ **do**
9 $\quad \lfloor$ $p(s_l, s_r) \leftarrow D_l[s_l] \times D_r[s_r]$;

10 Sample an (s_l, s_r) with probability $p(s_l, s_r)$ normalized by $\sum_{s_l, s_r} p(s_l, s_r)$;
11 $L \leftarrow$ UniformSampling(n_l, D, s_l);
12 $R \leftarrow$ UniformSampling(n_r, D, s_r);
13 **return** the tree rooted at n_i with child subtrees L and R;

---

rooted at a node n_i, we decide on the weight of the subtrees rooted at each child node, biased by the number of outcomes as counted in D, and then sample r-subtrees of the desired weights recursively.

We now explain Algorithm 3 in detail.

If $x = 0$ and $w(n_i) \neq 0$, we must return the empty tree (line 1 to line 2), since the empty tree is the only choice in this case. If $x = 0$ and $w(n_i) = 0$, there are $D_i[0]$ r-subtrees of weight 0 at node n_i, of which one is the empty tree while the others are non-empty retaining n_i. Therefore, to ensure the uniform distribution of the samples, we return the empty tree with probability $\frac{1}{D_i[0]}$ (line 4 to line 5), and return a non-empty tree of weight 0 retaining n_i with probability $1 - \frac{1}{D_i[0]}$ (the rest of the algorithm).

Except for the above two cases that return the empty subtree, we return n_i and subtrees t_l and t_r rooted at the children n_l and n_r of n_i. We first decide on the weight s_l and s_r of t_l and t_r (line 8 to line 10), biasing by the number of outcomes for each weight combination, and then recursively sample a subtree of the prescribed weight (line 11 to line 12), uniformly at random, and return it.

To be a suitable choice, the weight pair (s_l, s_r) must be such that $s_l + s_r = x - w(n_i)$ (which accounts for node n_i). Intuitively, to perform a uniform sampling, we now observe that the choice of the weight pair (s_l, s_r) partitions the set of outcomes. Hence, the probability that we select one weight pair should be proportional to the number of possible outcomes for this pair, namely, the number of r-subtrees t_l and t_r such that weight(t_l) = s_l and weight(t_r) = s_r. We compute this from D_l and D_r (line 9) by observing that the number of pairs (t_l, t_r) is the product of the number of choices for t_l and for t_r, as every combination of choices is possible.

Example 4. Follow Example 3. Assume we want to sample an r-subtree t' of weight$(t') = 3$ uniformly. Let D be the result of Algorithm 2.

We first call UniformSampling$(n_0, D, 3)$. We have to return n_0 (as $w(n_0) = 0$). Now n_0 has two children, n_1 and n_6. The only possible weight pair is $(0, 3)$, with probability $p(0, 3) = 1$. We now call recursively UniformSampling$(n_1, D, 0)$ and UniformSampling$(n_6, D, 3)$.

When callingUniformSampling$(n_1, D, 0)$, since $w(n_1) = 0$, we return $\emptyset$ with probability $\frac{1}{D_1[0]} = \frac{1}{2}$ and return n_1 with probability $1 - \frac{1}{D_1[0]} = \frac{1}{2}$. Assume n_1 is returned.

We proceed to UniformSampling$(n_6, D, 3)$. We have to return n_6 (note that $w(n_6) = 2$). Now n_6 has two children, n_2 and n_3. The possible weight pairs for this call are $(1, 0)$ and $(0, 1)$, with respective (unnormalized) probabilities $p(1, 0) = D_2[1] \times D_3[0] = 1 \times 1 = 1$ and $p(0, 1) = D_2[0] \times D_3[1] = 2 \times 1 = 2$. The normalized probabilities are $p(1, 0) = \frac{1}{3}$ and $p(0, 1) = \frac{2}{3}$. Assume that we choose $(0, 1)$ with probability $\frac{2}{3}$. We now call recursively UniformSampling$(n_2, D, 0)$ and UniformSampling$(n_3, D, 1)$.

When calling UniformSampling$(n_2, D, 0)$ (note $w(n_2) = 0$), we return $\emptyset$ with probability $\frac{1}{D_2[0]} = \frac{1}{2}$ and return n_2 with probability $1 - \frac{1}{D_2[0]} = \frac{1}{2}$. Assume n_2 is returned.

We finish with UniformSampling$(n_3, D, 1)$. Node n_3 is selected.

Hence, the end result is the r-subtree whose nodes are $\{n_0, n_1, n_6, n_2, n_3\}$ (and whose edges can clearly be reconstituted in PTIME from t). This r-subtree is selected with the probability $\frac{1}{2} \times \frac{2}{3} \times \frac{1}{2} = \frac{1}{6}$. Indeed, recall that we know there are 6 r-subtrees of t of weight 3, according to $D_0[3] = 6$.   □

We now show the tractability and correctness of Algorithm 3, concluding the proof of Theorem 1.

Lemma 2. *For any tree t, node $n_i \in V(t)$ and integer $0 \leqslant x \leqslant$ weight(n_i), given D computed by Algorithm 2, UniformSampling(n_i, D, x) terminates in $O(nk)$ time (where n is the number of nodes in the given tree and k is the desired weight) and returns an r-subtree of weight x rooted at n_i, uniformly at random (i.e., solves the sampling problem for binary trees).*

Proof. We first prove the complexity claim. On every node n_i of the binary tree t, the number of possibilities to consider is at most k, and for each possibility the number of operations performed is constant (assuming that drawing a number uniformly at random can be performed in constant time). The overall running time is $O(nk)$.

We now show correctness by induction on n_i. The base case is $n_i = $ NULL, in which case we must have $x = 0$ and we correctly return $\emptyset$. We now assume that n_i is not NULL. If $x = 0$ and $w(n_i) \neq 0$, the only possibility is the empty subtree and we correctly return $\emptyset$. If $x = 0$ and $w(n_i) = 0$, there are $D_i[0]$ r-subtrees of weight 0 at node n_i, of which one is the empty subtree while the others are non-empty and retain n_i. To ensure that the distribution is uniform, we return the empty tree with probability $\frac{1}{D_i[0]}$ and return a non-empty tree of weight 0 retaining n_i with probability $1 - \frac{1}{D_i[0]}$. Otherwise, we need to return an r-subtree retaining n_i. As in the proof of Lemma 1, the set of possible outcomes of the sampling process is partitioned by the possible assignments, and only the valid ones correspond to a non-empty set of outcomes. Hence, we can first choose a weight pair, *weighted by the proportion of outcomes which are outcomes for this pair*, and then choose an outcome for this pair. Now, observe that, by Lemma 1, D correctly represents the number of outcomes for each child of n_i, so that our computation of p (which mimics that of Algorithm 2) correctly represents the proportion of outcomes for each weight pair. We then choose an assignment according to p, and then observe that choosing an outcome for this assignment amounts to choosing an outcome for each child of n_i whose weight is given by the assignment. By induction hypothesis, this is precisely what the recursive calls to UniformSampling(n_i, D, x) perform. This concludes the proof. □

5.2 Sampling for Unranked Trees

In this section, we show that the algorithm of the previous section can be adapted so that it works on arbitrary unranked trees, not just binary trees.

We first observe that the straightforward generalization of Algorithm 1 to trees of arbitrary arity, where assignments and convolutions are performed for all children, is still correct. However, its running time would no longer be polynomial in n and k, as there would be a potentially exponential number of weight assignments to consider.

Fortunately, there is still hope to avoid considering all weights assignments over all the children, because *convolution is associative*. Informally, assuming we have three children $\{n_1, n_2, n_3\}$, we do the following: we treat $\{n_1\}$ as a group and $\{n_2, n_3\}$ as the second group, then enumerate weight pairs over $\{n_1\}$ and $\{n_2, n_3\}$; once a weight pair, in which a positive integer is assigned to $\{n_2, n_3\}$, is selected, we can treat $\{n_2\}$ and $\{n_3\}$ as new groups and enumerate weight pairs over $\{n_2\}$ and $\{n_3\}$. In other words, this strategy can be implemented by transforming the original tree to a binary tree.

We now present an encoding process transforming unranked trees to *encoded trees*, which are binary trees whose nodes are either regular nodes or *dummy* nodes. Intuitively, the encoding operation replaces sequences of more than two children by a hierarchy of dummy nodes representing those children; replacing dummy nodes by the sequence of their children yields back the original tree. The encoding is illustrated in Fig. 1, where the tree in Fig. 1(b) is the encoded tree of the one in Fig. 1(a) (dummy nodes are represented as squares). We require that the weight of every dummy node is 0, i.e., $\mathsf{w}(n_i) = 0$ where n_i is a dummy node. (However, dummy nodes are not exactly equivalent to regular nodes with weight 0, as we must not consider that we have the possibility of either keeping them or not keeping them.)

We formally present an algorithm (Algorithm 4) for this encoding process. Algorithm 4 performs a constant number of operations on every considered node plus a constant number of operations on every child of the considered node. Hence, the overall number of operations performed for every node (both when

---

Algorithm 4. Transforming to binary tree

Data: a tree t with nodes $n_0, \ldots, n_{k-1}$
Result: the encoded tree t'

1 $m \leftarrow k$ and $t' \leftarrow t$;
2 **for** *each node n_i of t* **do**
3   **if** $|\text{children}(n_i)| > 2$ **then**
4   create dummy nodes $n_m, n_{m+1}, \ldots, n_{m+|\text{children}(n_i)|-3}$ in $\mathsf{V}(t')$;
5   **for** $0 \leqslant j < |\text{children}(n_i)| - 3$ **do**
6   create an edge from n_{m+j} to n_{m+j+1} in t';
7   create an edge from n_i to n_m in t';
8   disconnect the 2nd, 3rd, $\ldots$, children of n_i from n_i in t';
9   **for** $0 \leqslant j \leqslant |\text{children}(n_i)| - 3$ **do**
10   create an edge in t' from n_{m+j} to the $(j+2)^{th}$ child of n_i in t;
11   create an edge in t' from $n_{m+|\text{children}(n_i)|-3}$ to the last child of n_i in t;
12       $m \leftarrow m + |\text{children}(n_i)| - 3 + 1$;

---

examining it and when examining its unique parent) is constant, so it completes in linear time. For a tree t with n nodes, the number of nodes in its encoded tree t' is no more than $n + n - 3$ (the worst case being achieved by a tree with a root node and $n - 1$ children). Therefore the size of the encoded tree is linear in the size of the original tree. Note that the weights of created dummy nodes are set to 0.

Based on this encoding process, we now state our result:

Theorem 2. *The sampling problem can be solved in $O(nk^2)$ (where n is the number of nodes in the given tree and k is the desired weight), for arbitrary unranked trees.*

Proof. It can be shown that, up to the question of keeping or deleting the dummy nodes with no regular descendants (we call them *bottommost*), there is a bijection between r-subtrees in the original tree and r-subtrees in the encoded tree. Hence, we can solve the sampling problem by choosing an r-subtree in the encoded tree with a set of regular nodes of weight k, uniformly at random, and imposing the choice of keeping bottommost dummy nodes.

We do this by adapting Algorithms 2 and 3 to run correctly on encoded trees, that is, managing dummy nodes correctly, by imposing that they are always retained.

In Algorithm 2, we have to define the computation of D_i for a dummy node n_i as $D_i \leftarrow D_l * D_r$ (as it must always be kept, and does not increase the weight of the r-subtree).

In Algorithm 3, some operations have to be distinguished between regular nodes and dummy nodes. In line 1 and line 4, we additionally require in the if clauses that n_i is either NULL or a regular node: indeed, for dummy nodes, even if $x = 0$ we cannot return $\emptyset$, because we must keep them.

The correctness and running time of the modified algorithms can be proved by straightforward adaptations of Lemmas 1 and 2. □

6 Tractable Uniform Sampling

As presented in the previous section, the complexity of Algorithm 1 (and its variant in Theorem 2) is $O(nk^2)$, where n is the number of nodes in the given tree and k is the desired weight. It is thus a *pseudo-polynomial time* algorithm to solve the general sampling problem, which is polynomial in the *value* of k, but is still exponential in the *size* of k.

We now observe that for the tractable cases in Sect. 4.2, Algorithm 1 (and its variant in Theorem 2) runs in time polynomial in the size of the input tree; more specifically, in $O(n^3)$. Indeed, for unweighted sampling (where $\mathsf{w}(n_i) = 1$ for every n_i) and 0/1-weights sampling (where $\mathsf{w}(n_i) = \{0, 1\}$ for every n_i), the desired weight k is bounded by the size of the tree, since $k \leqslant \mathsf{weight}(t) \leqslant n$. Therefore the complexity of Algorithm 1 (and its variant in Theorem 2) is then $O(n^3)$, that is, cubic in the size of the input tree. Hence, we have the following claim:

Theorem 3. *The unweighted sampling and 0/1-weights sampling can be solved in $O(n^3)$ time, where n is the number of nodes in the given tree.*

7 Repeated Requests

In this section, we consider the more general problem where the data consumer requests a *completion* of a certain price to data that they have already bought. The motivation is that, after having bought incomplete data, the user may realize that they need additional data, in which case they would like to obtain more incomplete data that is not redundant with what they already have.

A first way to formalize the problem is as follows, where data is priced according to a known subtree (provided by the data consumer) by considering that known nodes are free (but that they may or may not be returned again).

Definition 6. *The problem of sampling an r-subtree of weight k in a tree t conditionally to an r-subtree t' is to sample an r-subtree t'' of t uniformly at random, such that $\mathsf{weight}(t'') - \sum_{n \in (V(t') \cap V(t''))} \mathsf{w}(n) = k$.*

An alternative is to consider that we want to sample an *extension* of a fixed size to the whole subtree, so that all known nodes are always part of the output:

Definition 7. *The problem of sampling an r-subtree of weight k in a tree t that extends an r-subtree t' is to sample an r-subtree t'' of t uniformly at random, such that (1) t' is an r-subtree of t''; (2) $\mathsf{weight}(t'') - \mathsf{weight}(t') = k$.*

Note that those two formulations are not the same: the first one does not require the known part of the document to be returned, while the second one does. While it may be argued that the resulting outcomes are essentially equivalent (as they only differ on parts of the data that are already known to the data consumer), it is important to observe that they define different distributions: though both problems require the sampling to be uniform among their set of outcomes, the additional possible outcomes of the first definition means that the underlying distribution is not the same.

As the uniform sampling problem for r-subtrees can be reduced to either problem by setting t' to be the empty subtree, the NP-hardness of those two problems follows from Proposition 1. However, we can show that, in the unweighted case, those problems are tractable, because they reduce to the 0/1-weights sampling problem which is tractable by Theorem 3:

Proposition 2. *The problem of sampling an r-subtree of weight k in a tree t conditionally to an r-subtree t' can be solved in $O(n^3)$ time if t is unweighted. The same holds for the problem of sampling an r-subtree that extends another r-subtree.*

Proof. For the problem of Definition 6, set the weight of the nodes of t' in t to be zero (the intuition is that all the known nodes are free). The problem can then be solved by applying Theorem 2.

For the problem of Definition 7, set the weight of the nodes of t' in t to be zero but we have to ensure that the nodes in t' are always returned. To do so, we adapt Theorem 2 by handling the nodes in t' in the same way as handling dummy nodes in the previous section.   □

8   Sampling Extension: Sampling on Weight and Height

In this section, we consider a more complicated sampling scenario: sampling an r-subtree of weight k and height h uniformly at random. We present a pseudo-polynomial time algorithm for this sampling problem. To start with, we define the height of a tree.

Definition 8. *(Height of a tree) For a node $n \in V(t)$, we define inductively* $\mathsf{height}(n) := 1 + \max_{(n,n') \in E(t)} \mathsf{height}(n')$, *with* $\mathsf{height}(n) = 1$ *if n is a leaf of t. With slight abuse of notation, we note* $\mathsf{height}(t) = \mathsf{height}(\mathsf{root}(t))$ *the height of t.*

We first revisit the problem of pricing a tree depending on its weight and height. Then, in Sect. 8.2, we first describe the algorithm for the case of *binary* trees. Next, we adapt the algorithm in Sect. 8.3 to show how to apply it to arbitrary trees.

8.1   Pricing Function

Height, as well as weight, can be used as a measure of data completeness: two trees of same weight could be priced differently if they have a different height.

Algorithm 5. Algorithm for the sampling problem on binary trees

---

Input: a binary tree t and two integers $k \geqslant 0$ and $h \geqslant 0$
Result: an r-subtree t' of t of weight$(t') = k$ and height$(t') = h$ uniformly at
    random

// Phase 1: count the number of subtrees

1 $D \leftarrow$ SubtreeCounting(t);

// Phase 2: sample a random subtree

2 **if** $k \leqslant$ weight$(t) \wedge h \leqslant$ height$(t) \wedge D_{\text{root}(t)}[k, h] \neq 0$ **then**

3 | **return** UniformSampling$(\text{root}(t), D, k, h)$;

4 **else**

5 | fail ;

We are thus in a setting where a pricing function $\varphi_t(k, h)$ for the tree t should take into account both the weight k and the height h of an r-subtree. A healthy data market should impose at least the following conditions on φ_t:

Non-decreasing for weight. $k_1 \geqslant k_2 \Rightarrow \varphi_t(k_1, h) \geqslant \varphi_t(k_2, h)$.
Non-decreasing for height. $h_1 \geqslant h_2 \Rightarrow \varphi_t(k, h_1) \geqslant \varphi_t(k, h_2)$.
Arbitrage-free for weight. $\varphi_t(k_1, h) + \varphi_t(k_2, h) \geqslant \varphi_t(k_1 + k_2, h)$.
Arbitrage-free for both. $\varphi_t(k_1, h_1) + \varphi_t(k_2, h_2) \geqslant \varphi_t(k_1 + k_2, h_1 + h_2)$.

An example pricing function is $\varphi_t(k, h) := \alpha h + \beta \frac{k}{k+1}$, where $\alpha \geqslant 1 \geqslant \beta \geqslant 0$.

Similarly as in Sect. 3, once such a pricing function is fixed, the problem becomes to sample a tree of prescribed weight and height uniformly at random.

8.2 Sampling for Binary Trees

In this section, we provide an algorithm which proves the following theorem for binary trees:

Theorem 4. *The sampling problem for binary trees is solvable in time* $O(nk^2h^2)$, *where n is the number of nodes in the tree, k is the desired weight value, and h is the desired height.*

The sampling algorithm is adapted from the one in Sect. 5.1, and presented as Algorithm 5. The detailed adaptation of the two phases in the sampling algorithm is discussed in the following.

Modifications in Phase 1. As in Algorithm 2 (where $D_i[x]$ stores the number of r-subtrees of weight x rooted at node n_i), we need to record not only the weight but also the height of such r-subtrees. More precisely, we use $D_i[x, y]$ to denote the number of r-subtrees of weight x and height y rooted at node n_i, where $x \in [0, \text{weight}(t)]$ and $y \in [0, \text{height}(t)]$. We present Algorithm 6, to compute $D_i[x, y]$ for each node n_i in the given tree t.

Algorithm 6. SubtreeCounting(t)

Input: a binary tree t
Result: a matrix D such that $D_i[x, y]$ is the number of r-subtrees of weight x
and height y rooted at n_i for all n_i, x and y

1   **for** $x \in [0, \text{weight}(t)], y \in [0, \text{height}(t)], n_i \in V(t) \sqcup \{\text{NULL}\}$ **do**
2 | $D_i[x, y] \leftarrow 0$;
3    $D_{\text{NULL}}[0, 0] \leftarrow 1$;
    // We browse all nodes in topological order, from leaves to the root
4   **foreach** *non*-NULL *node n_i accessed bottom-up* **do**
5 | $D_i[0, 0] = 1$;
6 | $n_l \leftarrow$ first child of n_i;
7 | $n_r \leftarrow$ second child of n_i;
8 | $T \leftarrow D_l * D_r$;
9 | **for** $x \in [\text{w}(n_i), \text{weight}(n_i)]$ **do**
10 | **for** $y \in [1, \text{height}(n_i)]$ **do**
11 | | $D_i[x, y] \leftarrow T[x - \text{w}(n_i), y - 1]$;
12 **return** D;

As the base case, for NULL nodes, $\text{weight}(\text{NULL}) = 0$ and $\text{height}(\text{NULL}) = 0$. Hence $D_{\text{NULL}}[0, 0] = 1$ (as shown in line 3). For other cases (i.e., $x > 0$ or $y > 0$), $D_{\text{NULL}} = 0$.

For a non-NULL node, if the height is 0, then either the weight is also 0 and only the empty tree is possible, or the weight is greater than 0 and there is no possibility. Otherwise, intuitively, an r-subtree of weight $x \geqslant 0$ and height $y > 0$ rooted at node n_i is obtained by retaining n_i and choosing two r-subtrees t_l and t_r, respectively rooted at n_l and n_r (the children of n_i), such that $\text{weight}(t_l) + \text{weight}(t_r) = x - \text{w}(n_i)$ and $\max\{\text{height}(t_l), \text{height}(t_r)\} = y - 1$ (which accounts for the weight and height of the additional node n_i). Similar to Algorithm 2, the number of such choices is computed in line 8 as the convolution of D_l and D_r in a certain sense, defined as follows, for $0 \leqslant p \leqslant \text{weight}(t)$ and $0 \leqslant q \leqslant \text{height}(t)$:

$$(D_l * D_r)[p, q] := \sum_{\substack{0 \leqslant h_1, h_2 \leqslant q \\ h_1 = q \text{ or } h_2 = q}} \sum_{m=0}^{p} D_l[m, h_1] \times D_r[p - m, h_2]$$

$(D_l * D_r)[p, q]$ represents the number of pairs of r-subtrees t_l and t_r such that $\text{weight}(n_l) + \text{weight}(n_r) = p$ and $\max\{\text{height}(t_l), \text{height}(t_r)\} = q$. In other words, there are three mutually exclusive ways to meet the requirement on the heights:

1. $\text{height}(t_l) < q$ and $\text{height}(t_r) = q$;
2. $\text{height}(t_l) = q$ and $\text{height}(t_r) < q$;
3. $\text{height}(t_l) = q$ and $\text{height}(t_r) = q$.

All in all, as shown in line 9 to line 11, the number of r-subtrees of weight x and height y, namely, $D_i[x, y]$, is the number of pairs of r-subtrees t_l and t_r rooted at n_l and n_r respectively such that $\text{weight}(n_l) + \text{weight}(n_r) = x - \text{w}(n_i)$

Algorithm 7. UniformSampling(n_i, D, x, y)

Input: a node n_i (or NULL), the precomputed D, a weight value x and a height value y

Result: an r-subtree of weight x and height y at node n_i if one exists

1 **if** $y = 0$ **then**
2 | **return** $\emptyset$;

3 $n_l \leftarrow$ first child of n_i;
4 $n_r \leftarrow$ second child of n_i;
5 **for** $0 \leqslant s_l, s_r \leqslant x$ *and* $0 \leqslant o_l, o_r \leqslant y$ **s.t.** $s_l + s_r = x - \mathsf{w}(n_i)$ **and**
     $\max\{o_l, o_r\} = y - 1$ **do**
6 | $p([s_l, o_l], [s_r, o_r]) \leftarrow D_l[s_l, o_l] \times D_r[s_r, o_r]$;

7 Sample an $([s_l, o_l], [s_r, o_r])$ with probability $p([s_l, o_l], [s_r, o_r])$ normalized by
    $\sum_{([s_l, o_l], [s_r, o_r])} p([s_l, o_l], [s_r, o_r])$;
8 $L \leftarrow$ UniformSampling(n_l, D, s_l, o_l);
9 $R \leftarrow$ UniformSampling(n_r, D, s_r, o_r);
10 **return** the tree rooted at n_i with child subtrees L and R;

---

and $\max\{\mathsf{height}(t_l), \mathsf{height}(t_r)\} = y - 1$ (which is $T[x - \mathsf{w}(n_i), y - 1]$ in line 11). Note that when $\mathsf{w}(n_i) \neq 0$ there exists no such r-subtrees at node n_i of weight x (where $x \in [0, \mathsf{w}(n_i) - 1]$) and height $y \neq 0$, so $D_i[x, y]$ remains 0 in this case.

The time complexity to compute D is $O(nk^2h^2)$. To sample an r-subtree of weight k and height h, we need to record the number of r-subtrees of weight up to k and height up to h rooted at every node. Therefore each array in D is a $k \times h$ array. Computing the convolution sum of such two arrays takes $O(k^2h^2)$ time, since computing each value in the convolution sum takes $O(kh)$ time. The number of convolution sums to compute overall is $O(n)$, because each array D_i occurs in exactly one convolution sum. The overall running time is $O(nk^2h^2)$.

Modifications in Phase 2. Similarly to Algorithm 3, we present Algorithm 7 to sample an r-subtree of weight x and height y at node n_i uniformly at random, given the computed matrix D in the previous section. If $y = 0$, the only possible r-subtree is the empty tree, therefore that is the output (line 1 to line 2). Note that the result will be incorrect if $x > 0$, but in this case Algorithm 5 would not have called Algorithm 7 because there is no such subtree to sample.

Except for this special case, we return n_i and subtrees t_l and t_r rooted at the children n_l and n_r of n_i. We first decide on the weight (respectively, height) s_l (respectively, o_l) and s_r (respectively, o_r) of t_l and t_r (line 5 to line 7) before sampling recursively a subtree of the prescribed weight and the prescribed height (line 8 to line 9), uniformly at random, and returning it.

The possible weight and height pairs $([s_l, o_l], [s_r, o_r])$ must satisfy the following conditions to be possible choices for the weights and the heights of the subtrees t_l and t_r:

1. $0 \leqslant s_l, s_r \leqslant x$ and $0 \leqslant o_l, o_r \leqslant y$;
2. $s_l + s_r = x - \mathsf{w}(n_i)$ and $\max\{o_l, o_r\} = y - 1$ (which accounts for node n_i).

Intuitively, to perform a uniform sampling, we now observe that the choice of the weight and height pair $([s_l, o_l], [s_r, o_r])$ partitions the set of outcomes. Hence, the probability that we select one weight and height pair should be proportional to the number of possible outcomes for this pair, namely, the number of r-subtrees t_l and t_r such that $\mathsf{weight}(t_l) = s_l$, $\mathsf{weight}(t_r) = s_r$ and $\mathsf{height}(t_l) = o_l$, $\mathsf{height}(t_r) = o_r$. We compute this from D_l and D_r (line 6) by observing that the number of pairs (t_l, t_r) is the product of the number of choices for t_l and for t_r, as every combination of choices is possible.

The uniform sampling phase takes $O(nkh)$ time. On every node n_i of the binary tree t, the number of possibilities to consider is $O(kh)$ because every node has exactly two children, and for each possibility the number of operations performed is constant (assuming that drawing a number uniformly at random is constant-time). The overall running time is $O(nkh)$.

8.3   Sampling for Unranked Trees

In this section, we show that the algorithm of the previous section can be adapted so that it works on arbitrary unranked trees, not just binary trees. Similarly to Sect. 5.2, we transform an unranked tree to a binary tree whose nodes are either regular nodes or dummy nodes. A dummy node is a virtual node gathering a sequence of more than two nodes. Therefore a dummy node does not contribute any weight nor height to r-subtrees. After transforming an arbitrary unranked tree to the corresponding binary tree using Algorithm 4, we apply Algorithm 5 to solve the sampling problem, while making sure the dummy nodes are managed correctly. We explain how to adapt Algorithms 6 and 7 to handle the dummy nodes.

In Algorithm 6, we have to define the computation of D_i for a dummy node n_i as $D_i \leftarrow D_l * D_r$ (as it must always be kept, and does not increase the weight nor the height of the r-subtree).

In Algorithm 7, some operations have to be distinguished between regular nodes and dummy nodes. In line 1 we add one more condition in the if clause: n_i is either NULL or a regular node (for dummy nodes, even if $x = 0$ and $y = 0$ we cannot return $\emptyset$ as we must keep dummy nodes). In line 5, if n_i is a dummy node, the condition for possible weight and height pairs is: $s_l + s_r = x$ and $\max\{o_l, o_r\} = y$, because a dummy node does not contribute any weight nor height.

These adaptations do not affect the complexity of the sampling algorithm, therefore the algorithm for sampling unranked trees on both weight and height is also $O(nk^2h^2)$.

8.4   Tractable Cases

As presented in the previous section, the complexity of Algorithm 5 (and its variant in Sect. 8.3) is $O(nk^2h^2)$, where n is the number of nodes in the given tree, k is the desired weight and h is the desired height. As $h \leqslant n$, the time complexity of Algorithm 5 is $O(n^3k^2)$. It is a pseudo-polynomial time algorithm to solve the sampling problem on both weight and height, which is polynomial in the *value* of k, but is still exponential in the *size* of k.

The tractable cases in Sect. 4.2 are still tractable when we sample on both weight and height. To solve such tractable cases, Algorithm 5 (and its variant in Sect. 8.3) runs in polynomial-time to the size of the tree, more specifically, $O(n^5)$, because $k \leqslant n$ when $\mathsf{w}(n_i) = 1$ or $\mathsf{w}(n_i) \in \{0, 1\}$. Hence:

Theorem 5. *The unweighted sampling and 0/1-weights sampling on both weight and height can be solved in $O(n^5)$ time, where n is the number of nodes in the given tree.*

9   Conclusion

We proposed a framework for a data market in which data quality can be traded for a discount. We studied the case of XML documents with completeness as the quality dimension. Namely, a data provider offers an XML document, and sets both the price and weights of nodes of the document. The data consumer proposes a price but may get only a sample if the proposed price is lower than that of the entire document. A sample is a rooted subtree of prescribed weight, as determined by the proposed price, sampled uniformly at random.

We proved that if nodes in the XML document have arbitrary non-negative weights, the sampling problem is intractable. We devise a pseudo-polynomial time algorithm to solve this general sampling problem, and proved the time complexity and correctness of the algorithm. We identified tractable cases, namely the unweighted sampling problem and 0/1-weights sampling problem, for which the pseudo-polynomial time algorithm actually runs in polynomial time. We also considered repeated requests and provided PTIME solutions to the unweighted cases.

As a more complicated sampling scenario, we studied the problem of uniform sampling an r-subtree of prescribed weight and height. We devised a pseudo-polynomial time sampling algorithm, and showed that it still runs in polynomial time in the tractable cases.

The more general issue that we are currently investigating is that of sampling rooted subtrees uniformly at random under more expressive conditions than size restrictions or 0/1-weights (with or without height). In particular, we intend to identify the tractability boundary to describe the class of tree statistics for which it is possible to sample r-subtrees in PTIME under a uniform distribution.

Acknowledgments. This work is supported by the French Ministry of Foreign Affairs under the STIC-Asia program, CCIPX project.

References

1. Cohen, S., Kimelfeld, B., Sagiv, Y.: Running tree automata on probabilistic XML. In: PODS (2009)

2. Henzinger, M.R., Heydon, A., Mitzenmacher, M., Najork, M.: On near-uniform URL sampling. Comput. Netw. **33**(1–6), 295–308 (2000)
3. Hübler, C., Kriegel, H.-P., Borgwardt, K., Ghahramani, Z.: Metropolis algorithms for representative subgraph sampling. In: ICDM (2008)
4. Koutris, P., Upadhyaya, P., Balazinska, M., Howe, B., Suciu, D.: Query-based data pricing. In: PODS (2012)
5. Koutris, P., Upadhyaya, P., Balazinska, M., Howe, B., Suciu, D.: QueryMarket demonstration: pricing for online data markets. PVLDB **5**(12), 1962–1965 (2012)
6. Koutris, P., Upadhyaya, P., Balazinska, M., Howe, B., Suciu, D.: Toward practical query pricing with QueryMarket. In: SIGMOD (2013)
7. Leskovec, J., Faloutsos, C.: Sampling from large graphs. In: SIGKDD (2006)
8. Li, C., Li, D.Y., Miklau, G., Suciu, D.: A theory of pricing private data. In: ICDT (2013)
9. Li, C., Miklau, G.: Pricing aggregate queries in a data marketplace. In: WebDB (2012)
10. Lin, B.-R., Kifer, D.: On arbitrage-free pricing for general data queries. PVLDB **7**(9), 757–768 (2014)
11. Lu, X., Bressan, S.: Sampling connected induced subgraphs uniformly at random. In: Ailamaki, A., Bowers, S. (eds.) SSDBM 2012. LNCS, vol. 7338, pp. 195–212. Springer, Heidelberg (2012)
12. Luo, C., Jiang, Z., Hou, W.-C., Yu, F., Zhu, Q.: A sampling approach for XML query selectivity estimation. In: EDBT (2009)
13. Maiya, A.S., Berger-Wolf, T.Y.: Sampling community structure. In: WWW (2010)
14. Muschalle, A., Stahl, F., Löser, A., Vossen, G.: Pricing approaches for data markets. In: Castellanos, M., Dayal, U., Rundensteiner, E.A. (eds.) BIRTE 2012. LNBIP, vol. 154, pp. 129–144. Springer, Heidelberg (2013)
15. Pipino, L., Lee, Y.W., Wang, R.Y.: Data quality assessment. Commun. ACM **75**(4), 211–218 (2002)
16. Ribeiro, B.F., Towsley, D.F.: Estimating and sampling graphs with multidimensional random walks. In: Internet Measurement Conference (2010)
17. Tang, R., Amarilli, A., Senellart, P., Bressan, S.: Get a sample for a discount. In: Decker, H., Lhotská, L., Link, S., Spies, M., Wagner, R.R. (eds.) DEXA 2014, Part I. LNCS, vol. 8644, pp. 20–34. Springer, Heidelberg (2014)
18. Tang, R., Shao, D., Bressan, S., Valduriez, P.: What you pay for is what you get. In: Decker, H., Lhotská, L., Link, S., Basl, J., Tjoa, A.M. (eds.) DEXA 2013, Part II. LNCS, vol. 8056, pp. 395–409. Springer, Heidelberg (2013)
19. Wang, R.Y., Strong, D.M.: Beyond accuracy: what data quality means to data consumers. J. Manag. Inf. Syst. **12**(4), 5–33 (1996)
20. Wang, W., Jiang, H., Lu, H., Yu, J.X. Containment join size estimation: models and methods. In: SIGMOD (2003)

kdANN+: A Rapid AkNN Classifier for Big Data

Nikolaos Nodarakis[1](✉), Evaggelia Pitoura[2], Spyros Sioutas[3],
Athanasios Tsakalidis[1], Dimitrios Tsoumakos[3], and Giannis Tzimas[4]

[1] Computer Engineering and Informatics Department,
University of Patras, 26500 Patras, Greece
{nodarakis,tsak}@ceid.upatras.gr
[2] Computer Science Department, University of Ioannina, Ioannina, Greece
pitoura@cs.uoi.gr
[3] Department of Informatics, Ionian University, 49100 Corfu, Greece
{sioutas,dtsouma}@ionio.gr
[4] Computer and Informatics Engineering Department,
Technological Educational Institute of Western Greece, 26334 Patras, Greece
tzimas@cti.gr

Abstract. A k-nearest neighbor (kNN) query determines the k nearest points, using distance metrics, from a given location. An all k-nearest neighbor (AkNN) query constitutes a variation of a kNN query and retrieves the k nearest points for each point inside a database. Their main usage resonates in spatial databases and they consist the backbone of many location-based applications and not only. In this work, we propose a novel method for classifying multidimensional data using an AkNN algorithm in the MapReduce framework. Our approach exploits space decomposition techniques for processing the classification procedure in a parallel and distributed manner. To our knowledge, we are the first to study the kNN classification of multidimensional objects under this perspective. Through an extensive experimental evaluation we prove that our solution is efficient, robust and scalable in processing the given queries.

Keywords: Classification · Nearest neighbor · MapReduce · Hadoop · Multidimensional data · Query processing

1 Introduction

Classification is the problem of identifying to which of a set of categories a new observation belongs, on the basis of a training set of data containing observations (or instances) whose category membership is known. One of the algorithms for data classification uses the kNN approach [10]. It computes the k nearest neighbors (belonging to the training dataset) of a new object and classifies it to the category that belongs the majority of its neighbors.

© Springer-Verlag Berlin Heidelberg 2016
A. Hameurlain et al. (Eds.): TLDKS XXIV, LNCS 9510, pp. 139–168, 2016.
DOI: 10.1007/978-3-662-49214-7_5

A k-nearest neighbor query [19] computes the k nearest points, using distance metrics, from a specific location and is an operation that is widely used in spatial databases. An all k-nearest neighbor query constitutes a variation of a kNN query and retrieves the k nearest points for each point inside a dataset in a single query process. There is a wide diversity of applications that AkNN queries can be harnessed. The classification problem is one of them. Furthermore, they are widely used by location based services [13]. For example, consider users that send their location to a web server to process a request using a position anonymization system in order to protect their privacy from insidious acts. This anonymization system may use an AkNN algorithm to calculate the k nearest neighbors for each user. After that, it sends to the server the locations of the neighbors along with the location of the user that made the request at the first place. In addition, many algorithms have been developed to optimize and speed up the join process in databases using the kNN approach.

Although AkNN is a fundamental query type, it is computationally very expensive. The naive approach is to search for every point the whole dataset in order to estimate its k-NN list. This leads to an $O\left(n^2\right)$ time complexity assuming that n is the cardinality of the dataset. As a result, quite a few centralized algorithms and structures (M-trees, R-trees, space-filling curves, etc.) have been developed towards this direction [6,12,15,31]. However, as the volume of datasets grows rapidly even these algorithms cannot cope with the computational burden produced by an AkNN query process. Consequently, high scalable implementations are required. Cloud computing technologies provide tools and infrastructure to create such solutions and manage the input data in a distributed way among multiple servers. The most popular and notably efficient tool is the *MapReduce* [9] programming model, developed by Google, for processing large-scale data.

In this paper, we propose a method for efficient multidimensional data classification using AkNN queries in a single batch-based process in *Hadoop* [22,25], the open source MapReduce implementation. The basic idea is to decompose the space, where the data belongs, into smaller partitions. Afterwards, we get the k nearest neighbors for each point to be classified only by searching the appropriate partitions. Finally, we add it to the category it belongs based on the class that the majority of its neighbors belongs. The space decomposition relies on the data distribution of the training dataset.

More specifically, we sum up the technical contributions of our paper as follows:

- We present an implementation of a classification algorithm based on AkNN queries using MapReduce. We apply space decomposition techniques (based on data distribution) that divides the data into smaller groups. For each point we search for candidate k-NN objects only in a few groups. The granularity of the decomposition is a key factor for the performance of the algorithm and we analyze it further in Sect. 6.1. At first, the algorithm defines a search area for each point and investigates for k-NN points in the groups covered by this area. If the search area of a point does not include at least k neighbors, it is gradually expanded until the desired number is reached. Finally, we classify the point to

the category that belongs the majority of its neighbors. The implementation defines the MapReduce jobs with no modifications to the original Hadoop framework.
- We provide an extension for $d > 3$ in Sect. 5 (d stands for dimensionality).
- We evaluate our solution through an experimental evaluation against large scale data up to 4 dimensions. Furthermore, we study various parameters that can affect the total computational cost of our method using real and synthetic datasets. The results prove that our solution is efficient, robust and scalable.

The rest of the paper is organized as follows: Sect. 2 discusses related work. Section 3 presents the initial idea of the algorithm, our technical contributions and some examples of how the algorithm works. Section 4 presents a detailed analysis of the classification process developed in Hadoop. Section 5 provides an extension for $d > 3$ and Sect. 6 presents the experiments that where conducted in the context of this work. Finally, Sect. 7 concludes the paper and Sect. 8 presents future steps.

2   Related Work

AkNN queries have been extensively studied in literature. In [15], a method based on M-trees is proposed that processes AkNN spatial network queries. The experimental evaluation runs over a road network dataset for small k values. In addition, a structure that is popular for answering efficiently to kNN queries is R-tree [19]. Assuming that we execute a kNN query for all elements stored in the R-tree, we facilitate the AkNN query process with such indexes. Pruning techniques can be combined with such structures to deliver better results [6,12]. Mobile networks are also a domain where AkNN find application as shown in [4]. Their work suggest a centralized algorithm that identifies to every smartphone user its k geographically nearest neighbors in $O(n \cdot (k + l))$ time, where n denotes the number of users and l is a network-specific parameter. Moreover, efforts have been made to design low computational cost methods that execute such queries in spatial databases. For instance, [27] studies both the kNN query and the kNN join in a relational database. Their approach guarantees to find the approximate kNN with only logarithmic number of page accesses in expectation with a constant approximation ratio. Also, it can be extended to find the exact kNN efficiently in any fixed dimension. The works in [26,29] propose algorithms to answer kNN join.

The methods proposed above can handle data of small size in one or more dimensions, thus their use is limited in centralized environments only. During the recent years, the researchers have focused on developing approaches that are applicable in distributed environments, like our method, and can manipulate big data in an efficient manner. The MapReduce framework seems to be suitable for processing such queries. For example, in [28] the discussed approach splits the target space in smaller cells and looks into appropriate cells where k-NN objects are located, but applies only in 2-dimensional data. Our method speeds up the naive solution of [28] by eliminating the merging step, as it is a major drawback. We have to denote here that in [28] it is claimed that the computation of the merging step can be performed in one node since we just consider statistic values. But this is not entirely true since this process can derive a notable computational burden as we increase dimensions and/or data size, something that is confirmed in the experimental evaluation. In addition, the merging step can produce sizeable groups of points, especially as k increments, that can overload the first step of the AkNN process. Moreover, our method applies for more dimensions. Especially, for $d >= 3$ the multidimensional extension is not straightforward at all.

In [21], locality sensitive hashing (LSH) is used together with a MapReduce implementation for processing kNN queries over large multidimensional datasets. This solution suggests an approximate algorithm like the work in [30] (H-zkNNJ) but we focus on exact processing of AkNN queries. Furthermore, AkNN queries are utilized along with MapReduce to speed up and optimize the join process over different datasets [2,17] or support non-equi joins [24]. Moreover, [3] makes use of a R-tree based method to process kNN joins efficiently. Together with kNN, many other popular spatial queries have been studied and implemented efficiently on top of Hadoop/HBase frameworks [1,11,16].

In [5] a minimum spanning tree based classification model is introduced and it can be viewed as an intermediate model between the traditional k-nearest neighbor method and cluster based classification method. Another approach presented in [14] recommends parallel implementation methods of several classification algorithms, including k-nearest neighbor, bayesian model, decision tree. However, it does not contemplate neither the perspective of dimensionality nor parameter k.

In brief, our proposed method implemented in the Hadoop MapReduce framework, extends the traditional kNN classification algorithm and processes exact AkNN queries over massive multidimensional data. In this way, we achieve to classify a huge amount of objects in a single batch-based process. Compared to the aforementioned solutions, our method does not focus solely on the join operator but provides a more generalized framework to process AkNN queries. In other words, we boost the performance of the AkNN query process regardless the context of use of the query (kNN join, AkNN classification, etc.) The experimental evaluation considers a wide diversity of factors that can affect the execution time such as the value of k, the granularity of space decomposition, dimensionality and data distribution.

3   Overview of Classification Algorithm

In this section, we first define some notation and provide some definitions used throughout this paper. Table 1 lists the symbols and their meanings. Next, we outline the architecture of MapReduce model. Finally, we give a brief review of the method our solution relies on and then we extend it for more dimensions and tackle some performance issues.

Table 1. Symbols and their meanings

n	Granularity of space decomposition
k	Number of nearest neighbors
d	Dimensionality
D	A d-dimensional metric space
$dist(r,s)$	The distance from r to s
$kNN(r,S)$	The k nearest neighbors of r from S
$AkNNC(R,S)$	$\forall r \in R$ classify r based on $kNN(r,S)$
$ICCH$	Interval, cell cube or hypercube
$ICSH$	Interval, circle, sphere or hypersphere
I	Input dataset
T	Training dataset
c_r	The class of point r
C_T	The set of classes of dataset T
S_I	Size of input dataset
S_T	Size of training dataset
M	Total number of Map tasks
R	Total number of Reduce tasks

3.1 Definitions

We consider points in a d-dimensional metric space D. Given two points r and s we define as $dist(r,s)$ the distance between r and s in D. In this paper, we used the distance measure of Euclidean distance

$$(r,s) = \sqrt{\sum_{i=1}^{d} (r[i] - s[i])^2}$$

where $r[i]$ (respectively $s[i]$) denote the value of r (respectively s) along the i-th dimension in D. Without loss of generality, alternative distance measures (i.e. Manhattan distance) can be applied to our solution.

Definition 1. kNN: *Given a point r, a dataset S and an integer k, the k nearest neighbors of r from S, denoted as $kNN(r,S)$, is a set of k points from S such that $\forall p \in kNN(r,S)$, $\forall q \in \{S - kNN(r,S)\}, dist(p,r) < dist(q,r)$.*

Definition 2. AkNN: *Given two datasets R, S and an integer k, the all k nearest neighbors of R from S, named $AkNN(R,S)$, is a set of pairs (r,s) such that $AkNN(R,S) = \{(r,s) : r \in R, s \in kNN(r,S)\}$.*

Definition 3. AkNN Classification: *Given two datasets R, S and a set of classes C_S where points of S belong, the classification process produces a set of pairs (r, c_r), denoted as $AkNNC(R,S)$, such that $AkNNC(R,S) = \{(r, c_r) : r \in R, c_r \in C_S\}$ where c_r is the class where the majority of $kNN(r,S)$ belong $\forall r \in R$.*

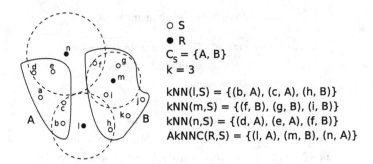

Fig. 1. AkNNC(R,S) explanation

We explain Definition 3 using an illustrative example, as shown in Fig. 1. We assume that $S = \{a, b, c, d, e, f, g, h, i, j, k\}, R = \{l, m, n\}, C_S = \{A, B\}$ and $k = 3$. We draw the boundary circle (see below in Sect. 3.2) that covers at least k points and construct $kNN(r, S), \forall r \in R$. Next, we determine the dominant class c_r in each $kNN(r, S), \forall r \in R$ and build the final $AkNNC(R, S)$ set.

3.2   Classification Using Space Decomposition

Consider a training dataset T, an input dataset I and a set of classes C_T where points of T belong. First of all, we define as *target space* the space enclosing the points of I and T. The partitions that are defined when we decompose the target space for 1-dimensional objects are called *intervals*. Respectively, we call *cells* and *cubes* the partitions in case of 2 and 3-dimensional objects and hypercubes for $d > 3$. For a new $1D$ point p, we define as *boundary interval* the minimum interval centred at p that covers at least k-NN elements. Respectively, we define the *boundary circle* and *boundary sphere* for $2D$ and $3D$ points and the *boundary hypersphere* for $d > 3$. The notion of hypercube and hypersphere are analyzed further in Sect. 5. When the boundary ICSH centred in an ICCH $icch_1$, intersects the bounds of an other $icch_2$ we say an *overlap* occurs on $icch_2$. Finally, for a point $i \in I$, we define as *updates* of $kNN(i, T)$ the existence of many different instances of $kNN(i, T)$ that need to be unified to a final set.

We place the objects of T on the target space according to their coordinates. The main idea of equal-sized space decomposition is to partition the target space into n^d equal sized ICCHs where n and the size of each ICCH are user defined. Each ICCH contains a number of points of T. Moreover, we construct a new layer over the target space according to C_T and $\forall t \in T, c_t \in C_T$. In order to estimate $AkNNC(I, T)$, we investigate $\forall i \in I$ for k-nearest neighbors only in a few ICCHs, thus bounding the number of computations that need to be performed for each i.

3.3   MapReduce Model

Here, we briefly describe the MapReduce model [9]. The data processing in MapReduce is based on input data partitioning; the partitioned data is executed by a number of tasks executed in many distributed nodes. There exist two major task categories called *Map* and *Reduce* respectively. Given input data, a *Map* function processes the data and outputs key-value pairs. Based on the Shuffle process, key-value pairs are grouped and then each group is sent to the corresponding Reduce task. A user can define his own Map and Reduce functions depending on the purpose of his application. The input and output formats of these functions are simplified as key-value pairs. Using this generic interface, the user can focus on his own problem and does not have to care how the program is executed over the distributed nodes. The architecture of MapReduce model is depicted in Fig. 2.

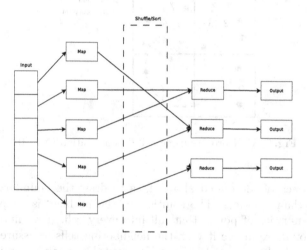

Fig. 2. Architecture of MapReduce model

3.4   Previous Work

A very preliminary study of naïve AkNN solutions is presented in [28] and uses a simple cell decomposition technique to process AkNN queries on two different datasets, i.e. I and T. The objects consisting both datasets are 2-dimensional points having only one attribute, the coordinate vector and the target space comprises of $2^n \times 2^n$ equal-sized cells.

The elements of both datasets are placed on the target space according to their coordinate vector and a cell decomposition is applied. For a point $i \in I$ it is expected that its $kNN(i, T)$ will be located in a close range area defined by nearby cells. At first, we look for candidate k-NN points inside the cell that i belongs in the first place, name it cl. If we find at least k elements we draw the boundary circle. There is a chance the boundary circle centred at cl overlaps some neighboring cells. In this case, we need to investigate for possible k-NN objects inside these overlapped cells in order to create the final k-NN list. If no overlap occurs, the k-NN list of i is complete. Next, we present an example to provide a better perception of the algorithm.

Figure 3 illustrates an example of the AkNN process of a point in dataset I using a query for $k = 3$. Initially, the point looks for k-NN objects inside cell 2. Since there exist at least 3 points of dataset T in cell 2 the boundary circle can be drawn. The boundary circle overlaps cells 1, 3 and 4, so we need to investigate for additional k-NN objects inside them. The algorithm outputs an instance of the k-NN list for every overlapped cell. These instances need to be unified into a k-NN list containing the final points (x, y, z).

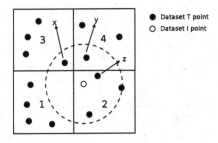

Fig. 3. kNN process using cell decomposition ($k = 3$)

This approach, as described above, fails to draw the boundary circle if cl contains less than k points. The solution to the problem is simple. At first, we check the number of points that fall into every cell. If we find a cell with less than k points we merge it with the neighboring cells to assure that it will contain the required number of objects. The way the merging step is performed relies on the principles of hierarchical space decomposition used in quad-trees [20]. Note that this is the reason why the space decomposition involves $2^n \times 2^n$ cells. This imposes two more steps that need to be done before we begin calculating $kNN(i, T)$. In the beginning, a counting phase needs to be performed followed by a merging step in order to overcome the issue mentioned above. This preprocessing phase induces additional cost to the total computation and, as shown in the experiments, the merging step can lead to a bad algorithmic behavior.

3.5 Technical Contributions

In this subsection, we extend the previous method for more dimensions and adapt it to the needs of the classification problem. Moreover, we analyze some drawbacks of the method studied in [28] and propose a mechanism to make the algorithm more efficient.

Firstly, we have a training dataset T, an input dataset I and a set of classes C_T where points of T belong. The only difference now is that the points in the training dataset have one more attribute, the class they belong. In order to compute $AkNNC(I, T)$, a classification step is executed after the construction of the k-NN lists. The class of every new object is chosen based on the class membership of its k-nearest neighbors. Furthermore, now the space is decomposed in 2^{dn} ICCHs since we consider a d-dimensional metric space D.

As mentioned before, the simple solution presented in [28] has one major drawback which is the merging step. Figure 4(a) depicts a situation where the merging step of the original method can significantly increase the total cost of the algorithm. Consider two points x and y entering cells 3 and 2 respectively and $k = 3$. We can draw point's x boundary circle since cell 3 includes at least k elements. On the contrary, we cannot draw the boundary circle of point y, so we need to unify cells 1 through 4 into one bigger cell. Now point y can draw its boundary circle but we overload point's x k-NN list construction with redundant computations. In the first place, the k-NN list of point x would only need 4 distance calculations to be formed. After the merging step we need to perform 15, namely almost 4 times more than before and this would happen for all points that would join cells 1,3 and 4 in the first place.

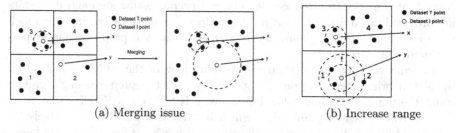

(a) Merging issue   (b) Increase range

Fig. 4. Issue of the merging step before the kNN process and way to avoid it ($k = 3$)

In order to avoid a scenario like above, we introduce a mechanism where only points that cannot find at least k-nearest neighbors in the ICCH in the first place proceed to further actions. Let a point p joining an ICCH $icch$ that encloses $l < k$ neighbors. Instead of performing a merging step, we draw the boundary ICSH based on these l neighbors. Then, we check if the boundary ICSH overlaps any neighboring ICCHs. In case it does, we investigate if the boundary ICSH covers at least k elements in total. In case it does, then we are able to build the final k-NN list of the point by unifying the individual k-NN lists that are derived for every overlapped ICCH. In case the boundary ICSH does not cover at least k objects in total or does not overlap any ICCHs, then we gradually increase its search range (by a fraction of the size of the ICCH each time) until the prerequisites are fulfilled.

Figure 4(b) explains this issue. Consider two points x and y entering cells 3 and 1 respectively and $k = 3$. We observe that cell 3 contains 4 neighbors and point x can draw its boundary circle that covers k-NN elements. However, the boundary circle centred at point y does not cover k-NN elements in the first place. Consequently, we gradually increase its search range until the boundary circle encloses at least k-NN points. Note that eliminating the merging step, we also relax the condition of decomposing the target space into 2^{dn} equal-sized splits and generalize it to n^d equal-sized splits.

Summing up, our solution can be implemented as a series of MapReduce jobs as shown below. These MapReduce jobs will be analyzed in detail in Sect. 4:

1. **Distribution Information.** Count the number of points of T that fall into each ICCH.
2. **Primitive Computation Phase.** Calculate possible k-NN points $\forall i \in I$ from T in the same ICCH.
3. **Update Lists.** Draw the boundary ICSH $\forall i \in I$ and increase it, if needed, until it covers at least k-NN points of T. Check for overlaps of neighboring ICCHs and derive updates of k-NN lists.
4. **Unify Lists.** Unify the updates of every k-NN list into one final k-NN list $\forall i \in I$.
5. **Classification.** Classify all points of I.

In Fig. 5, we illustrate the working flow of the AkNN classification process. Note, that the first MapReduce job acts as a preprocessing step and its results are provided as additional input in MapReduce Job 3 (to determine how much we need to increase the boundary ICSH) and that the preprocessing step is executed only once for T.

In order to fully comprehend the working flow of the AkNN classification process, a brief interpretation of Fig. 5 follows. In the preprocessing step, we count the number of points that fall in every ICCH (e.g. ICCH 2 contains 13 points). Now, consider points A, C which belong to ICCH $2, 3$ respectively. In the second MapReduce job, we derive an initial k-NN list for points A, C based on the objects contained in the same ICCH. In MapReduce job 3, observe that the boundary ICSH of A overlaps ICCH $3, 4$ and a new k-NN list instance is produced for each of them. The flag in each record is *false*, which indicates the need of extra computations to build the k-NN list. On the other hand, the flag of C equals to *true* and its k-NN list does not need any amendment. In MapReduce job 4, the two k-NN lists of A are unified to a final one and lastly, in the fifth MapReduce job we classify A, C based on the majority of class membership of their neighbors.

4   AkNN Classification with MapReduce

In this section, we present a detailed description of the classification process as implemented in the Hadoop framework. The whole process consists of five MapReduce jobs which are divided into three phases. Phase one estimates the distribution of T over the target space. Phase two determines $kNN(i, T), \forall i \in I$ and phase three estimates $AkNNC(I, T)$. The records in T have the format

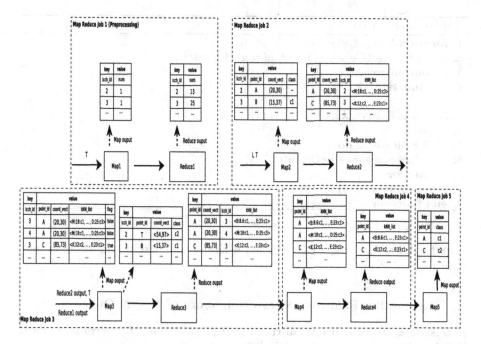

Fig. 5. Overview of the AkNN classification process

<point_id, coordinate_vector, class> and in I have the format <point_id, coordinate_vector>. Furthermore, parameters n and k are defined by the user. In the following subsections, we describe each MapReduce job separately and analyze the Map and Reduce functions that take place in each one of them. For each MapReduce job, we also quote pseudo-code, in order to provide a better comprehension of the Map and Reduce functions, and proceed to time and space complexity analysis.

4.1 Getting Distribution Information of Training Dataset

This MapReduce job is a preprocessing step required by subsequent MapReduce jobs that receive its output as additional data. In this step, we decompose the entire target space and count the number of points of T that fall in each ICCH. Below, we sum up the Map and Reduce functions of this MapReduce process.

The *Map* function takes as input records with the training dataset format. Afterwards, it estimates the ICCH id for each point based on its coordinates and outputs a key-value pair where the key is ICCH id and the value is number 1. The *Reduce* function receives the key-value pairs from the Map function and for each ICCH id it outputs the number of points of T that belong to it.

Each Map task needs $O\left(S_T/M\right)$ time to run. Each Reduce task needs $O\left(n^d/R\right)$ time to run as the total number of ICCHs is n^d. So, the size of the

---

MapReduce Job 1.

```
 1: function MAP(k1, v1)
 2:     coord = getCoord(v1); icch_id = getId(coord);
 3:     output(icch_id, 1);
 4: end function

 5: function REDUCE(k2, v2)
 6:     sum = 0;
 7:     for all v ∈ v2 do
 8:         sum = sum + getSum(v);
 9:     end for
10:     output(k2, sum);
11: end function
```

---

output will be $O\left(n^d \cdot c_{\text{si}}\right)$, where c_{si} is the size of sum and icch_id for an output record.

4.2 Estimating Primitive Phase Neighbors of AkNN Query

In this stage, we concentrate all training (L_T) and input (L_I) records for each ICCH and compute possible k-NN points for each item in L_I from L_T inside the ICCH. Below, we condense the Map and Reduce functions. We use two Map functions, one for each dataset, as seen in MapReduce Job 2 pseudo-code.

For each point $t \in T$, *Map1* outputs a new key-value pair in which the ICCH id where t belongs is the key and the value consists of the id, coordinate vector and class of t. Similarly, for each point $i \in I$, *Map2* outputs a new key-value pair in which the ICCH id where i belongs is the key and the value consists of the id and coordinate vector of i. The *Reduce* function receives a set of records from both Map functions with the same ICCH ids and separates points of T from points of I into two lists, L_T and L_I respectively. Then, the Reduce function calculates the distance for each point in L_I from L_T. Subsequently, it estimates the k-NN points and forms a list L with the format $< p_1, d_1, c_1 : \ldots : p_k, d_k, c_k >$, where p_i is the i-th NN point, d_i is its distance and c_i is its class. Finally, for each $p \in L_I$, *Reduce* outputs a new key-value pair in which the key is the id of p and the values comprises of the coordinate vector, ICCH id and list L of p.

Each Map1 task needs $O\left(S_T/M\right)$ time and each Map2 task needs $O\left(S_I/M\right)$ time to run. For a Reduce task, suppose u_i and t_i the number of input and training points that are enclosed in an ICCH in the i-th execution of a Reduce function and $1 \leq i \leq n^d/R$. The Reduce task needs $O\left(\sum_i u_i \cdot t_i\right)$. Let L_s to be the size of k-NN list and icch_id $\forall i \in I$. The output size is $O\left(S_I \cdot L_s\right) = O\left(S_I\right)$.

4.3 Checking for Overlaps and Updating k-NN Lists

In this step, at first we gradually increase the boundary ICSH (how much depends on information from the first MapReduce job), where necessary, until it

MapReduce Job 2.

```
 1: function MAP1(k1, v1)
 2:     coord = getCoord(v1); p_id = getPointId(v1);
 3:     class = getClass(v1); icch_id = getId(coord);
 4:     output(icch_id, < p_id, coord, class >);
 5: end function

 6: function MAP2(k1, v1)
 7:     coord = getCoord(v1);
 8:     p_id = getPointId(v1);
 9:     icch_id = getId(coord);
10:     output(icch_id, < p_id, coord >);
11: end function

12: function REDUCE(k2, v2)
13:     L_T = getTrainingPoints(v2);
14:     L_I = getInputPoints(v2);
15:     for all p ∈ L_I do
16:         L = List{};
17:         for all t ∈ L_T do
18:             L.add(new Record(t, dist(p, t), t.class));
19:         end for
20:         output(p.id, < p.coord, k2, getKNN(L) >);
21:     end for
22: end function
```

includes at least k points. Then, we check for overlaps between the ICSH and the neighboring ICCHs and derive updates of the k-NN lists. The Map and Reduce functions are outlined in MapReduce Job 3 pseudo-code. Again, we have two Map functions but the pseudo-code of Map1 function is omitted since it is the same with the respective function from MapReduce Job 2.

For each point $i \in I$, function *Map2* computes the overlaps between the ICSH and the neighboring ICCHs. If no overlap occurs, it does not need to perform any additional steps. It outputs a key-value pair in which ICCH id is the key and the value consists of id, coordinate vector and list L of i and a flag *true* which implies that no further process is required. Otherwise, for every overlapped ICCH it outputs a new record where ICCH id' (id of an overlapped ICCH) is the key and the value consists of id, coordinate vector and list L of i and a flag *false*. The flag indicates we need to search for possible k-NN objects inside the overlapped ICCHs. The *Reduce* function receives a set of points with the same ICCH ids and separates the points of T from points of I into two lists, L_T and L_I respectively. After that, the Reduce function performs extra distance calculations using the points in L_T and updates k-NN lists for the records in L_I. Finally, for each $p \in L_I$ it generates a record in which the key is the id of p and the values comprises of the coordinate vector, ICCH id and list L of p.

MapReduce Job 3.

```
 1: function MAP2(k1, v1)
 2:     c = getCoord(v1); p_id = getPointId(v1);
 3:     kNN = getKNNList(v1); r = getRadius(kNN);
 4:     while kNN.size() < k do
 5:         increase(r); kNN.addAll(getNeighbors(r));
 6:     end while
 7:     oICCHs = getOverlappedICCHs(r);
 8:     if oICCHs.size() > 0 then
 9:         for all icch ∈ oICCHs do
10:             output(icch, < p_id, c, kNN, false >);
11:         end for
12:     else
13:         output(getId(c), < p_id, c, kNN, true >);
14:     end if
15: end function

16: function REDUCE(k2, v2)
17:     L_T = getTrainingPoints(v2); L_I = getInputPoints(v2);
18:     for all p ∈ I do
19:         if p.flag == true then
20:             output(p.id, < p.coord, key, p.kNN >);
21:         else
22:             L = List{};
23:             for all t ∈ T do
24:                 L.add(new Record(t, dist(p, t), t.class));
25:             end for
26:             L_f = finalKNN(L, p.kNN);
27:             output(p.id, < p.coord, key, L_f >);
28:         end if
29:     end for
30: end function
```

As before, each Map1 task needs $O\left(S_T/M\right)$ time to run. Consider an unclassified point p initially belonging to an ICCH $icch$. Let r be the number of times we increase the search range for p and $icchov$ the number of ICCHs that may be overlapped for p. For each Map2 task the i-th execution of the Map function performs $icchov_i + r_i$ steps, where $1 \leq i \leq S_I/M$. So, each Map2 task runs in $O\left(\sum_i (icchov_i + r_i)\right)$ time. For a Reduce task, suppose u_i and t_i the number of points of I and T respectively that are enclosed in an ICCH in the i-th execution of a Reduce function and $1 \leq i \leq n^d/R$. The Reduce task needs $O\left(\sum_i u_i \cdot t_i\right)$. The size of updated records is a fraction of S_I. So, the size of the output is also $O\left(S_I\right)$.

4.4   Unifying Multiple k-NN Lists

The previous step it is possible to yield multiple updates of a point's k-NN list. This MapReduce job tackles this problem and unifies possible multiple lists into one final k-NN list for each point $i \in I$. The Map and Reduce functions are summarized at MapReduce Job 4 pseudo-code below.

MapReduce Job 4.

```
1: function MAP(k1, v1)
2:     output(getPointId(v1), getKKN(v1));
3: end function

4: function REDUCE(k2, v2)
5:     L = List{};
6:     for all v ∈ v2 do
7:         L.add(v);
8:     end for
9:     output(k2, unifyLists(L));
10: end function
```

The *Map* function receives the records of the previous step and extracts the k-NN list for each point. For each point $i \in I$, it outputs a key-value pair in which the key is the id of i and the value is the list L. The *Reduce* function receives as input key-value pairs with the same key and computes $kNN(i,T), \forall i \in I$. The key of an output record is again the id of i and the value consists of $kNN(i,T)$.

Each Map task runs in $O(S_I/M)$. For each Reduce task, assume $updates_i$ the number of updates for the k-NN list of an unclassified point in the i-th execution of a Reduce function, where $1 \leq i \leq |N_I|/R$ and $|N_I|$ the number of points in input dataset. Then, each Reduce task needs $O(\sum_i updates_i)$ to run. Let, I_{id} the size of ids of all points in I and L_{final} is the size of the final k-NN list $\forall i \in I$. The size of L_{final} is constant and I_{id} is $O(S_I)$. Consequently, the size of the output is $O(S_I)$.

4.5   Classifying Points

This is the final job of the whole classification process. It is a Map-only job that classifies the input points based on the class membership of their k-NN points. The Map function receives as input records from the previous job and outputs $AkNNC(I,T)$. More precisely, each record handled by the *Map* function is a point together with a list of class occurrences of its k-NN neighbors. The function parses iteratively the list and reports the class with the highest cardinality. The key of an output record is the id of the point given as input to the Map function, while the value is the class the point is assigned. Each Map task runs in $O(S_I/M)$ time and output size is $O(S_I)$.

MapReduce Job 5.

```
1: function MAP(k1, v1)
2:      H = HashMap < Class, Occurences > {};
3:      H = findClassOccur(v1);
4:      max = 0; maxClass = null;
5:      for all entry ∈ H do
6:          if entry.occur > max then
7:              max = entry.occur;
8:              maxClass = entry.class;
9:          end if
10:     end for
11:     output(getPointId(v1), maxClass);
12: end function
```

5   Extension for $d > 3$

Here we provide the extension of our method for $d > 3$. In geometry, a hypercube [7,8] is a n-dimensional analogue of a square ($n = 2$) and a cube ($n = 3$) and is also called a n-cube (i.e. 0-cube is a hypercube of dimension zero and represents a point). It is a closed, compact and convex figure that consists of groups of opposite parallel line segments aligned in each of the space's dimensions, perpendicular to each other and of the same length.

Respectively, an n-sphere [7,8] is a generalization of the surface of an ordinary sphere to a n-dimensional space. Spheres of dimension $n > 2$ are called hyperspheres. For any natural number n, an n-sphere of radius r is defined as a set of points in $(n + 1)$-dimensional Euclidean space which are at distance r from a central point and r may be any positive real number. So, the n-sphere centred at the origin is defined by:

$$S^n = \{x \in \Re^{n+1} : \| x \| = r\}$$

Figure 6 displays how to create a hypercube for $d = 4$ (4-cube) from a cube for $d = 3$. Regarding our solution for $d > 3$, the target space now is decomposed into equal-sized d-dimensional hypercubes and in the first place we investigate for k-NN points in each hypercube. Next, we draw the boundary hypersphere and increase it, if needed, until it bounds at least k neighbors. Afterwards, we inspect for any overlaps between the boundary hypersphere and neighboring hypercubes. Finally, we build the final k-NN list for each unclassified point and categorize it according to class majority of its k-NN list.

6   Experimental Evaluation

In this section, we conduct a series of experiments to evaluate the performance of our method under many different perspectives. More precisely, we take into consideration the value of k, granularity of space decomposition, dimensionality and data distribution.

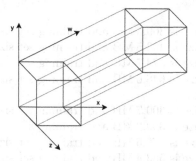

Fig. 6. Creating a 4-cube from a 3-cube

Our cluster includes 32 computing nodes (VMs), each one of which has four 2.1 GHz CPU processors, 4 GB of memory, 40 GB hard disk and the nodes are connected by 1 GB Ethernet. On each node, we install Ubuntu 12.04 operating system, Java 1.7.0_40 with a 64-bit Server VM, and Hadoop 1.0.4. To adapt the Hadoop environment to our application, we apply the following changes to the default Hadoop configurations: the replication factor is set to 1; the maximum number of Map and Reduce tasks in each node is set to 3, the DFS chunk size is 256 MB and the size of virtual memory for each Map and Reduce task is set to 512 MB.

We evaluate the following approaches in the experiments:

– kdANN is the solution proposed in [28] along with the extension (which invented and implemented by us) for more dimensions, as described in Sect. 3, in order to be able to compare it with our solution.
– kdANN+ is our solution for d-dimensional points without the merging step as described in Sect. 3.

We evaluate our solution using both real[1] and synthetic datasets. We create 1D and 2D datasets from the real dataset keeping the x and the (x, y) coordinates respectively. In addition, by using statistics of underlying real dataset, we add one more dimension z in order to construct a 4-dimensional dataset. We process the datasets to fit into our solution (i.e. normalization) and we end up with 1D, 2D, 3D and 4D datasets that consist of approximately 19,000,000 points and follow a power law like distribution. From each dataset, we extract a fraction of points (10 %) that are used as a training dataset. Respectively, we create 1, 2, 3 and 4-dimensional datasets with uniformly distributed points, each dataset has 19,000,000 points and the training datasets contain 1,900,000 points. For each point in a training dataset we assign a class based on its coordinate vector. The file sizes of datasets are:

---

[1] The real dataset is part of the Canadian Planetary Emulation Terrain 3D Mapping Dataset, which is a collection of 3-dimensional laser scans gathered at two unique planetary analogue rover test facilities in Canada. The dataset provides the coordinates (x, y, z) for each laser scan in meters. http://asrl.utias.utoronto.ca/datasets/ 3dmap/.

1. Real Dataset
    (a) $1D$: Input set size is 309.5 MB and training set size is 35 MB
    (b) $2D$: Input set size is 403.5 MB and training set size is 44.2 MB
    (c) $3D$: Input set size is 523.7 MB and training set size is 56.2 MB
    (d) $4D$: Input set size is 648.6 MB and training set size is 67.4 MB
2. Synthetic Dataset
    (a) $1D$: Input set size is 300.7 MB and training set size is 33.9 MB
    (b) $2D$: Input set size is 359.2 MB and training set size is 39.8 MB
    (c) $3D$: Input set size is 478.5 MB and training set size is 51.7 MB
    (d) $4D$: Input set size is 583.4 MB and training set size is 60.9 MB

We run experiments for data up to four dimensions due to the curse of dimensionality. As shown in the experiments, for $d > 2$ the total execution cost rises exponentially and for $d > 4$ overcomes the computational power of our cluster infrastructure. We can dodge such limitations by incorporating in our system dimensionality reduction techniques, such as Principal Component Analysis (PCA) or Singular Value Decomposition [18], or elasticity mechanisms [23]. We leave this kind of extension for future work, as stated in Sect. 8, since it is beyond the scope of this paper.

6.1   Tuning Parameter n

One major aspect in the performance of the algorithm is the tuning of granularity parameter n. In this experiment, we explain how to select a value of n in order to succeed in achieving the shortest execution time. Each time the target space is decomposed into 2^{dn} equal partitions in order for kdANN to be able to perform the merging step, as described in Sect. 3.

In the case of power law distributions, we choose higher values of n compared to uniform distributions. The intuition behind this idea, is that we want to discretize the target space into splits that contain as few points as possible in order to avoid an overload of the primitive computation phase. On the other hand, as n increases, the number of update steps also increases. This can overwhelm the AkNN process if the number of derived instances of the k-NN lists is massive. Regarding uniform distributions, we wish to create larger partitions, but again not too big, in order to avoid executing many update steps. Each time, the selection of n depends on the infrastructure of the cluster.

In Fig. 7, we depict how execution time varies as we alter value n in case of 2-dimensional real dataset for $k = 5$. In case of kdANN+, we notice that as parameter n grows the execution time drops and achieves its lowest value for $n = 9$ and slightly increases for $n = 10$. In contrary, the execution time for kdANN increases until $n = 9$ and drops significantly for $n = 10$. Moreover, its lowest achieved value is almost ten times bigger than kdANN+. Considering the above, we deduce that for power law distributions kdANN+ outperforms kdANN as n changes. In addition, we conclude that the merging step affects greatly the performance of kdANN and creates a wide divergence in total running time as n mutates.

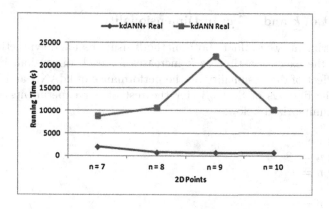

Fig. 7. Effect of n (real dataset $2D$)

Figure 8, presents the results of execution time for both methods when datasets follow a uniform distribution. Again, kdANN+ performs better than kdANN. Nevertheless, now the curve of running time presents a same behavior for both methods and when $n = 7$ the minimum running time is achieved. Observing the exported results from Figs. 7 and 8, we confirm our claim that we choose higher values of n in case of power law distribution datasets, compared to uniformly distributed datasets, in order to minimize the total execution time.

We proceed to similar experimental procedures for all dimensions. The results for $1D$, $3D$ and $4D$ points follow the same trend (we omit the graphs of other dimensions to avoid pointless repetition). In the case of real datasets, we pick value n that maximizes the performance of kdANN+ since kdANN presents a bad algorithmic behavior regardless of value n, as shown in the majority of experiments that follow.

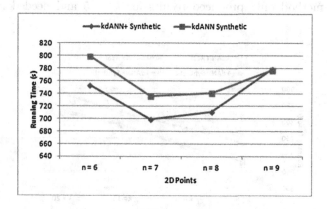

Fig. 8. Effect of n (synthetic dataset $2D$)

6.2   Effect of k and Effect of Dimensionality

In this experiment, we evaluate both methods using real and synthetic datasets
and record the execution time as k increases for each dimension. Finally, we
study the effect of dimensionality on the performance of kdANN and kdANN+.
Based on the findings of Sect. 6.1, for the rest of our experiments we set the
value n as summarized below:

1. Real Dataset
   (a) 1D: $n = 18$
   (b) 2D: $n = 9$
   (c) 3D: $n = 7$
   (d) 4D: $n = 6$
2. Synthetic Dataset
   (a) 1D: $n = 16$
   (b) 2D: $n = 7$
   (c) 3D: $n = 5$
   (d) 4D: $n = 4$

Effect of k for Different Dimensions. Figure 9 presents the results for kdANN
and kdANN+ by varying k from 5 to 20 on real and synthetic datasets. In terms
of running time, kdANN+ always perform better, followed by kdANN and each
method behave in the same way for both datasets, real and synthetic. As the
value of k grows, the size of each intermediate record becomes larger respectively.
Consequently, the data processing time increments. Moreover, as the number of
neighbors we need to estimate each time augments, we need to search into more
intervals for possible k-NN points as the boundary interval grows larger.

In Fig. 10, we demonstrate the outcome of the experimental procedure for
2-dimensional points when we alter k value from 5 to 20. First of all, note
that we do not include the results of kdANN for the real dataset. The reason
is that the method only produced results for $k = 5$ and needed more than

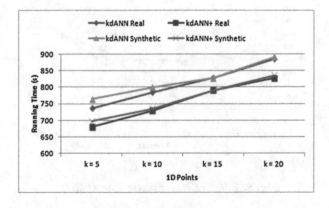

Fig. 9. Effect of k for $d = 1$

4 h. Beyond this, the merging step of kdANN derived extremely sizeable cells. Consequently, during the primitive computation phase a bottleneck was created to some nodes that strangled their resources, thus preventing them to yield any results. Observing the rest of the curves, we notice that the processing times are a bit higher than the previous ones due to larger records, as we impose one more dimension. Furthermore, the search area now overlaps more partitions of the target space than in case of 1-dimensional points. Consequently, the algorithm produces more instances of the k-NN lists and the time requirement to merge them is bigger. Overall, in the case of power law distribution, kdANN+ behaves much better than kdANN since the last one fails to process the AkNN query as k increases. Also, kdANN+ is faster and in case of synthetic dataset that follows a uniform distribution, especially as k increases.

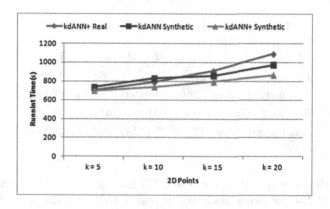

Fig. 10. Effect of k for $d = 2$

Figure 11 displays the results generated from kdANN and kdANN+ for the 3-dimensional points when we increase k value from 5 to 20. Once again, in case of kdANN we could not get any results for any value of k when we provided the real dataset as input. The reasons are the same we mentioned in the previous paragraph for $d = 2$.

Table 2 is pretty illustrative in the way the merging step affects the AkNN process. First of all, its computational cost is far from negligible if performed in a node (in contrary with the claim of the authors as stated in [28]). Apart from this, the ratio of cubes that participate in the merging process is almost 40 % and the largest merged cube consists of 32,768 and 262,144 initial cubes for $k = 5$ and $k > 5$ respectively.

In the case of kdANN+, when given the real dataset as input, it is obvious that the total computational cost is much larger compared to the one shown in Figs. 9 and 10. This happens for 3 reasons: (1) we have larger records in size, (2) some cubes are quite denser compared to others (since the dataset follows a power law distribution) and we need to perform more computations for them in the primitive computation phase and (3) a significant amount of overlaps

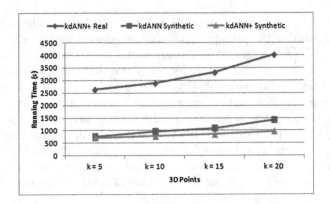

Fig. 11. Effect of k for $d = 3$

Table 2. Statistics of merging step for kdANN

	$k = 5$	$k = 10$	$k = 15$	$k = 20$
Time (s)	271	675	962	1,528
# of merged cubes	798,032	859,944	866,808	870,784
% of total cubes	38 %	41 %	41.3 %	41.5 %
Max merged cubes	32,768	262,144	262,144	262,144

take place, thus the update step of the k-NN lists needs more time than before. Finally, kdANN+ performs much better than kdANN, in the case of synthetic dataset, and the gap between the curves of running time tends to be bigger as k increases.

Finally, Fig. 12 demonstrates the total running cost for both kdANN and kdANN+ in the case of 4-dimensional datasets. Our method kdANN+, continues to overrun kdANN when our input follows a uniform distribution and the variance between the curves is a bit bigger than the previous cases. As expected, kdANN flunks in producing results for the real dataset while our method answers the AkNN query requiring much more processing time than the 3-dimensional case. The curve has a tendency to increase exponentially (we explained the reasons in the previous paragraph) and for $k = 20$ the time taken to export the outcome of the AkNN query is almost double compared to the running time of Fig. 11.

Effect of Dimensionality. In this subsection, we evaluate the effect of dimensionality for both real and synthetic datasets. Figure 13 presents the running time for $k = 20$ by varying the number of dimensions from 1 to 4.

From the outcome, we observe that kdANN is more sensitive to the number of dimensions than kdANN+ when we provide a dataset with uniform distribution as input. In particular, when the number of dimensions varies from 2 to 4 the

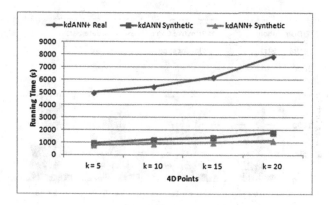

Fig. 12. Effect of k for $d = 4$

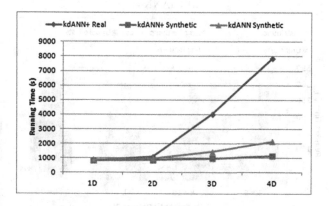

Fig. 13. Effect of dimensionality for $k = 20$

divergence between the two curves starts growing faster. In the case of power law distribution, we only include the results for kdANN+ since kdANN fails to process the AkNN query for dimensions 2 to 4 when $k = 20$. We notice that the execution time increases exponentially when $d > 2$. This results from the curse of dimensionality. As the number of dimensions increases, the number of distance computations as well as the number of searches in neighboring ICCHs increases exponentially. Nevertheless, kdANN+ can still process the AkNN query in a reasonable amount of time in contrast to kdANN.

6.3 Phase Breakdown

In Figs. 14(a)-14(c) we present the results of running time for different stages of kdANN and kdANN+, in case of 3-dimensional datasets, as k increases. We observe, that in all figures, the running time of distribution phase is the same (it runs only once since it is a preprocessing step). On the other hand, the running time of primitive computational and classification phase slightly increase as k

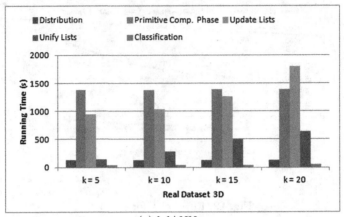

(a) kdANN+

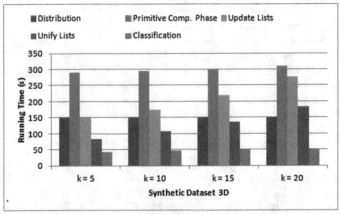

(b) kdANN+

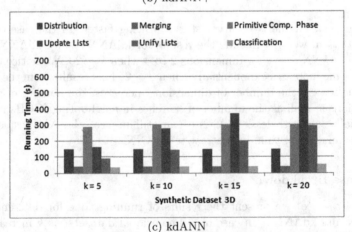

(c) kdANN

Fig. 14. Phase breakdown vs k (Color figure online).

grows. Since the k-NN list gets bigger, the algorithm takes more time to process the input records. Considering update and integrate phase, the running time increases notably. The bigger the value of k, the bigger the cardinality of derived instances of the k-NN lists due to larger area coverage by the boundary ICSH. Consequently, the algorithm needs more time to derive the final k-NN lists. The cumulative cost of these two phases is the one that mostly affects the total running time of the AkNN query in the majority of the experiments. Finally, the execution time of the merging phase remains the same (in case of kdANN in Fig. 14(c)). Apart from the merging phase, whose running cost may increase significantly (Table 2), the execution time for the rest phases follow the same trend as d varies.

6.4   Power Law vs Uniform Distribution

In this subsection, we perform a comparative analysis of the results exported by our method for datasets with different distributions and argue about the performance of methods kdANN and kdANN+ as k and d increments.

At first, we observe that as k increases kdANN+ prevails kdANN for all dimensions and for both dataset distributions (based on results from Figs. 9-12). It is clear that our contribution presented in Sect. 3.4, speeds up the solution presented in [28]. Under the perspective of dimensionality, in case of uniform distribution the divergence between the curves is not very big. Nevertheless, the running time of kdANN+ increases linearly whilst kdANN's running time grows exponentially for $d > 2$ (see Fig. 13). On the other hand, in case of power law distribution, for $d > 1$ kdANN+ outperforms kdANN. The last one either fails to derive results in a reasonable amount of time or cannot produce any results at all (again see Fig. 13). As shown in Table 2, the merging step has major deficiencies. It can cumber with notable computational burden the total AkNN process and can produce quite large merged ICCHs. As a consequence, the workload is badly distributed among the nodes and some of them end up running out of resources, thus causing kdANN to fail to produce any results. Despite the superiority of kdANN+, its execution time increases exponentially when the number of dimensions varies from 2 to 4.

Overall, the experimental evaluation shows that our solution (kdANN+) scales better than kdANN for uniform distributions and dominates it for power law distributions. However, it is clear that both kdANN and kdANN+ are more sensitive to power law distributions. As a result, their performance degrades faster than the case of uniform distributions.

6.5   Scalability

In this experiment, we investigate the scalability of the two approaches. We utilize the $3D$ datasets, since their size is quite big, and create new chunks smaller in size that are a fraction F of the original datasets, where $F \in \{0.2, 0.4, 0.6, 0.8\}$. Moreover, we set the value of k to 5.

Figure 15 presents the scalability results for real and synthetic datasets. In the case of power law distribution, the results display that kdANN+ scales almost linearly as the data size increases. In contrast, kdANN fails to generate any results even for very small datasets since the merging step continues to be an inhibitor factor in kdANN's performance. In addition, we can see that kdANN+ scales better than kdANN in the case of synthetic dataset and the running time increases almost linearly as in the case of power law distribution. Regarding kdANN, the curve of execution time is steeper until $F = 0.6$ and after that it increases more smoothly.

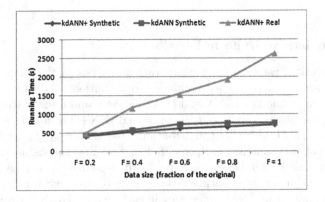

Fig. 15. Scalability

Table 3 shows the way the merging step affects kdANN as the data size varies. The ratio of cubes that are involved in the merging process remains high and varies from 36.6 % to 39.3 % and the largest merged cube comprises of 32,768 cubes of the initial space decomposition. Interestingly, the time to perform the merging step is not strictly increasing proportionally to the data size. In particular, the worst time is achieved when $F = 0.2$, then it reaches its minimum value for $F = 0.4$ and beyond this value augments again. Below, we explain why this phenomenon appears. The merging process takes into account the distribution information of dataset T. As the size of the input dataset decreases, respectively the size of the training dataset also mitigates. Since both datasets follow a power law distribution, the ICCHs that include training set points decrease also in number and this may result in more merging steps (i.e. $F = 0.2$).

6.6   Speedup

In our last experiment, we measure the effect of the number of computing nodes. We test four different cluster configurations and the cluster consist of $N \in \{11, 18, 25, 32\}$ nodes each time. We test the cluster configurations against the 3-dimensional datasets when $k = 5$.

Table 3. Statistics of merging step for kdANN and different data sizes

	$F = 0.2$	$F = 0.4$	$F = 0.6$	$F = 0.8$
Time (s)	598	223	279	300
# of merged cubes	825,264	767,768	768,256	802,216
% of total cubes	39.3 %	36.6 %	36.6 %	38.2 %
Max merged cubes	32,768	32,768	32,768	32,768

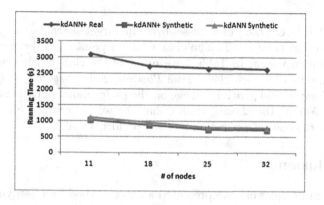

Fig. 16. Speedup

From Fig. 16, we observe that total running time of kdANN+, in the case of power law distribution, tends to decrease as we add more nodes to the cluster. Due to the increment of number of computing nodes, the amount of distance calculations and update steps on k-NN lists that undertakes each node decreases respectively. Moreover, since kdANN fails to produce any results using 3-dimensional real dataset when the cluster consists of 32 nodes, it is obvious that it will fail with less nodes too. That is the reason for the absence of kdANN's curve from Fig. 16. In the case of synthetic dataset, we observe that both kdANN and kdANN+ achieve almost the same speedup as the number of nodes increases; still kdANN+ performs betters than kdANN. We behold that in the case of real dataset the curve of running time decreases steeper as the number of nodes varies from 11 to 18 and becomes smoother beyond this point. On the other hand, in case of synthetic dataset the curves decrease smoother when the number of nodes varies from 25 to 32. The conclusion that accrues from this observation is that the increment of computing nodes has a greater effect on the running time of both approaches when the datasets follow a uniform distribution. This happens because the workload is distributed better among the nodes of the cluster.

6.7 Classification Performance

In this section, we present the performance results of our classification method for kdANN+, when the $3D$ real dataset is provided as input and $k = 10$. We define

Table 4. Classification performance of kdANN+

	Class A	Class B	Class C	Average
True positive	99.97%	99.91%	94.14%	98%
False negative	0.03%	0.09%	5.86%	2%
False positive	0.13%	0.06%	0.06%	0.08%
True negative	99.87%	99.94%	99.94%	99.92%

a set $C_T = \{A, B, C, D, E\}$ of 5 classes over the target space, but only 3 of them (A, B, C) contain points of T. The class where a point $t \in T$ belongs, depends on its coordinate vector. In Table 4, we measure the classification performance using four metrics for each class, *True Positive, False Negative, False Positive* and *True Negative* and give an average on the performance of each metric for all the classes. Among the classes, class C has the worst accuracy but the overall results show that our classification method performs well.

7 Conclusions

In the context of this work, we presented a novel method for classifying multidimensional data using AkNN queries in a single batch-based process in Hadoop. To our knowledge, it is the first time a MapReduce approach for classifying multidimensional data is discussed. By exploiting equal-sized space decomposition techniques we bound the number of distance calculations we need to perform for each point to reckon its k-nearest neighbors. We conduct a variety of experiments to test the efficiency of our method on both, real and synthetic datasets. Through this extensive experimental evaluation we prove that our system is efficient, robust and scalable.

8 Future Work

In the near future, we plan to extend and improve our system in order to boost its efficiency and flexibility. At first, we want to relax the condition of decomposing the target space into equal-sized splits. We have in mind to implement a technique that will allow us to have unequal splits, containing approximately the same number of points. This is going to decrease the number of overlaps and calculations for candidate k-NN points. Moreover, the method will become distribution independent leading to better load balancing between the nodes.

In addition, we intend to apply a mechanism in order for the cluster to be used in a more elastic way, by adding (respectively removing) nodes as the number of dimensions increase (respectively decrease) or the data distribution becomes more (respectively less) challenging to handle.

Finally, we plan to use indexes, such as R-trees or M-trees, along with HBase, in order to prune any points that are redundant and cumber additional cost to the method.

Acknowledgements. This research has been co-financed by the European Union (European Social Fund ESF) and Greek national funds through the Operational Program "Education and Lifelong Learning" of the National Strategic Reference Framework (NSRF) - Research Funding Program: Thales. Investing in knowledge society through the European Social Fund.

References

1. Aji, A., Wang, F., Vo, H., Lee, R., Liu, Q., Zhang, X., Saltz, J.: Hadoop GIS: a high performance spatial data warehousing system over MapReduce. Proc. VLDB Endow. **6**, 1009–1020 (2013)
2. Afrati, F.N., Ullman, J.D.: Optimizing joins in a map-reduce environment. In: Proceedings of the 13th International Conference on Extending Database Technology, pp. 99–110. ACM, New York (2010)
3. Böhm, C., Krebs, F.: The k-nearest neighbour join: turbo charging the KDD process. Knowl. Inf. Syst. **6**, 728–749 (2004)
4. Chatzimilioudis, G., Zeinalipour-Yazti, D., Lee, W.-C., Dikaiakos, M. D.: Continuous all k-nearest-neighbor querying in smartphone networks. In: Proceedings of the 2012 IEEE 13th International Conference on Mobile Data Management, pp. 79–88. IEEE Computer Society, Washington, DC (2012)
5. Chang, J., Luo, J., Huang, J.Z., Feng, S., Fan, J.: Minimum spanning tree based classification model for massive data with mapreduce implementation. In: Proceedings of the 10th IEEE International Conference on Data Mining Workshop, pp. 129–137. IEEE Computer Society, Washington, DC (2010)
6. Chen, Y., Patel, J.M.: Efficient evaluation of all-nearest-neighbor queries. In: Proceedings of the 23rd IEEE International Conference on Data Engineering, pp. 1056–1065. IEEE Computer Society, Washington, DC (2007)
7. Coxeter, H.S.M.: Regular Polytopes. Dover Publications, New York (1973)
8. Cromwell, P.R.: Polyhedra. Cambridge University Press, Cambridge (1999)
9. Dean, J., Ghemawat, S.: MapReduce: simplified data processing on large clusters. In: Proceedings of the 6th Symposium on Operating Systems Design and Implementation, pp. 137–150. USENIX Association, Berkeley (2004)
10. Dunham, M.H.: Data Mining: Introductory and Advanced Topics. Prentice Hall, Upper Saddle River (2002)
11. Eldawy, A.: SpatialHadoop: towards flexible and scalable spatial processing using mapreduce. In: Proceedings of the 2014 SIGMOD Ph.D. Symposium, pp. 46–50. ACM, New York (2014)
12. Emrich, T., Graf, F., Kriegel, H.-P., Schubert, M., Thoma, M.: Optimizing all-nearest-neighbor queries with trigonometric pruning. In: Gertz, M., Ludäscher, B. (eds.) SSDBM 2010. LNCS, vol. 6187, pp. 501–518. Springer, Heidelberg (2010)
13. Gkoulalas-Divanis, A., Verykios, V.S., Bozanis, P.: A network aware privacy model for online requests in trajectory data. Data Knowl. Eng. **68**, 431–452 (2009)
14. He, Q., Zhuang, F., Li, J., Shi, Z.: Parallel implementation of classification algorithms based on mapreduce. In: Yu, J., Greco, S., Lingras, P., Wang, G., Skowron, A. (eds.) RSKT 2010. LNCS, vol. 6401, pp. 655–662. Springer, Heidelberg (2010)
15. Ioup, E., Shaw, K., Sample, J., Abdelguerfi, M.: Efficient AKNN spatial network queries using the m-tree. In: Proceedings of the 15th Annual ACM International Symposium on Advances in Geographic Information Systems, pp. 46:1–46:4. ACM, New York (2007)

16. Lee, K., Ganti, R.K., Srivatsa, M., Liu, L.: Efficient spatial query processing for big data. In: Proceedings of the 22nd ACM SIGSPATIAL International Conference on Advances in Geographic Information Systems, pp. 469–472. ACM, New York (2014)
17. Lu, W., Shen, Y., Chen, S., Ooi, B.C.: Efficient processing of k nearest neighbor joins using mapreduce. Proc. VLDB Endow. **5**, 1016–1027 (2012)
18. Rajaraman, A., Ullman, J.D.: Mining of Massive Datasets. Cambridge University Press, New York (2011)
19. Roussopoulos, N., Kelley, S., Vincent, F.: Nearest neighbor queries. In: Proceedings of the 1995 ACM SIGMOD International Conference on Management of Data, pp. 71–79. ACM, New York (1995)
20. Samet, H.: The quadtree and related hierarchical data structures. ACM Comput. Surv. **16**, 187–260 (1984)
21. Stupar, A., Michel, S., Schenkel, R.: RankReduce - processing k-nearest neighbor queries on top of MapReduce. In: Proceedings of the 8th Workshop on Large-Scale Distributed Systems for Information Retrieval, pp. 13–18 (2010)
22. The apache software foundation: Hadoop homepage. http://hadoop.apache.org/
23. Tsoumakos, D., Konstantinou, I., Boumpouka, C., Sioutas, S., Koziris, N.: Automated, elastic resource provisioning for nosql clusters using tiramola. In: Proceedings of the 13th IEEE/ACM International Symposium on Cluster, Cloud, and Grid Computing, pp. 34–41 (2013)
24. Vernica, R., Carey, M.J., Li, C.: Efficient parallel set-similarity joins using MapReduce. In: Proceedings of the ACM SIGMOD International Conference on Management of Data, pp. 495–506. ACM, New York (2010)
25. White, T.: Hadoop: The Definitive Guide, 3rd edn. O'Reilly Media/Yahoo Press (2012)
26. Xia, C., Lu, H., Chin, B., Hu, O.J.: Gorder: An efficient method for KNN join processing. In: VLDB, pp. 756–767. VLDB Endowment (2004)
27. Yao, B., Li, F., Kumar, P.: K nearest neighbor queries and KNN-joins in large relational databases (almost) for free. In: Proceedings of the 26th International Conference on Data Engineering, pp. 4–15. IEEE Computer Society, Washington, DC (2010)
28. Yokoyama, T., Ishikawa, Y., Suzuki, Y.: Processing all k-nearest neighbor queries in hadoop. In: Gao, H., Lim, L., Wang, W., Li, C., Chen, L. (eds.) WAIM 2012. LNCS, vol. 7418, pp. 346–351. Springer, Heidelberg (2012)
29. Yu, C., Cui, B., Wang, S., Su, J.: Efficient index-based KNN join processing for high-dimensional data. Inf. Softw. Technol. **49**, 332–344 (2007)
30. Zhang, C., Li, F., Jestes, J.: Efficient parallel kNN joins for large data in MapReduce. In: Proceedings of the 15th International Conference on Extending Database Technology, pp. 38–49. ACM, New York (2012)
31. Zhang, J., Mamoulis, N., Papadias, D., Tao, Y.: All-nearest-neighbors queries in spatial databases. In: Proceedings of the 16th International Conference on Scientific and Statistical Database Management, pp. 297–306. IEEE Computer Society, Washington, DC (2004)

Optimizing Inter-data-center Large-Scale Database Parallel Replication with Workload-Driven Partitioning

Zhen Gao[3(✉)], Hong Min[1(✉)], Xiao Li[2], Jie Huang[3], Yi Jin[4], An Lei[3],
Serge Bourbonnais[2], Miao Zheng[5], and Gene Fuh[5]

[1] IBM T. J. Watson Research Center, Yorktown Heights, NY, USA
hongmin@us.ibm.com
[2] IBM Silicon Valley Lab, San Jose, CA, USA
{lixi,bourbon}@us.ibm.com
[3] School of Software Engineering, Tongji University, Shanghai, China
{gaozhen,huangjie,1434318}@tongji.edu.cn
[4] Pivotal Inc., Beijing, China
jinyi.smilodon@gmail.com
[5] IBM System and Technology Group, New York, USA
zhengm@cn.ibm.com, fuh@us.ibm.com

Abstract. Geographically distributed data centers are deployed for non-stop business operations by many enterprises. In case of disastrous events, ongoing workloads must be failed over from the current data center to another active one within just a few seconds to achieve continuous service availability. Software-based parallel database replication techniques are designed to meet very high throughput with near-real-time latency. Understanding workload characteristics is one of the key factors for improving replication performance. In this paper, we propose a workload-driven method to optimize database replication latency and minimize transaction splits with a minimum of parallel replication consistency groups. Our two-phased approach includes (1) a log-based mechanism for workload pattern discovery; (2) a history-based algorithm on pattern analysis, database partitioning and partition adjustment. The experimental results from a real banking batch workload and a benchmark OLTP workload demonstrate the effectiveness of the solution even for partitioning 1000 s of database tables in very large workloads. Finally, the algorithm to automate the cyclic flow of workload profile capturing and partitioning readjustment is developed and verified.

1 Introduction

Many enterprises employ multiple geographically distributed data centers running the same applications and having the same data to provide zero-downtime upgrades, cross-site workload balancing, continuous availability and disaster recovery. Across these centers, data replication is used to maintain multiple data copies in near real-time.

Y. Jin—Work done while employed by IBM.

© Springer-Verlag Berlin Heidelberg 2016
A. Hameurlain et al. (Eds.): TLDKS XXIV, LNCS 9510, pp. 169–192, 2016.
DOI: 10.1007/978-3-662-49214-7_6

Various database replication techniques are proposed to serve different purposes. High availability (HA) within a single data center employs data replication to maintain global transaction consistency [3] or to improve fault tolerance and system performance via transaction processing localization [1, 16, 18]. Replication over unlimited distances leads to enormous challenges of scalability, efficiency and reliability, especially in an active-active deployment with heterogeneous database architectures.

Although DBMS built-in replication function has the potential for better performance via tighter software stack integration, middleware-based replication is more suitable for multi-vendor heterogeneous database environments [12]. Industrial examples of such technology include IBM Infosphere Data Replication [22], Oracle GoldenGate [23], etc. One widely used approach is to capture committed data changes from DBMS recovery log and to replicate the changes to target DBMS. Replicating data after changes committed at the source does not impact the response time of source-side applications. This paper addresses the performance problem of large-scale asynchronous database replication optimization for minimizing the data staleness and data loss in case of unrecoverable disasters.

Parallel replication is a desirable solution to increase the throughput by concurrently replicating changed data through multiple logical end-to-end replication channels. Such concurrent replication can potentially split a transaction's writeset among channels. Similar to DBMS snapshot consistency, point-in-time (PIT) snapshot consistency is provided via time-based coordination among replication channels [22]. PIT consistency is a guarantee of replicated data having a consistent view with the source view at an instance of past time. Such a time delay in PIT consistency is called PIT consistency latency. PIT consistency latency at the target DBMS is determined by both replication channel throughput and the duration between when the first element of a transaction's writeset is replicated and when the last element is replicated. It is not difficult to envision that higher replication throughput delivers lower PIT consistency latency. In addition, normally the more replication channels a transaction's writeset is split into, the longer it takes to reach PIT consistency. Over-provisioning with underutilized replication channels also introduces extra complexities and wastes resources.

This paper addresses the partitioning automation in parallel database replication cross data centers. Partitioning a database is a challenging task in PIT consistency latency reduction. By following design principles of DBMS data independence [2], databases and applications are often designed separately. Database access patterns usually differ by applications. One specific database partitioning scheme hardly suits the needs of other applications. Furthermore, new applications are continually deployed on existing databases and access patterns change as business requirements evolve. Taking into account of varying workload characteristics, scale of database objects and resource constraints, it is impractical for database administrators to have comprehensive understandings of all the database activities and to manually perform and adjust database partitioning for parallel data replications.

Our design aims at minimizing replication channels and achieving desired point-in-time consistency. With the observation that similar workload patterns re-occur in most business applications, we propose a two-stepped approach. In our approach, a log-based mechanism is employed for workload pattern discovery, and a history-based algorithm is used for pattern analysis and database partitioning. The partition granularity is at DBMS object level such as tables and table partitions, which can reach up to thousands or tens of thousands in a large enterprise IT environment. Finer grained partitioning, such as at the row level, is less practical due to higher overhead in runtime replication coordination and DBMS contention resolution. Our approach discovers and analyzes the data access patterns from the DBMS recovery log, and makes partitioning recommendations using a proposed two-phased algorithm called Replication Partition Advisor (RPA)-algorithm. In the first phase, the algorithm finds a partitioning solution with the least replication channels such that the PIT consistency latency is below a threshold tied to a service-level agreement (SLA). The second phase refines the partitioning solution to minimize the number of transaction splits. Our approach is applicable to share-nothing, share-memory and share-disk databases [20]. The real-world workload evaluation and analysis demonstrate the effectiveness of our solution.

The rest of the paper is organized as follows. Section 2 introduces more background about inter-data-center parallel data replication. Section 3 describes the workload profile tool (WPT). Section 4 presents the RPA-algorithm and discusses how a real-time re-partitioning can be achieved. The experiment evaluations are presented in Sect. 5. We discuss related work in Sect. 6 and end the paper with the conclusion in Sect. 7.

2 Background on Parallel Data Replication

Based on the data change propagation, mainstream replication technologies can be classified into two major types: synchronous replication (also called eager replication) and asynchronous replication (also called lazy replication). In eager replication, the changes to all the copies are in a single transaction (unit of work). If a failure happens on any copy, the entire transaction will roll back. Compared with eager replication, lazy replication eliminates the impact on transaction response time by relaxing the strong consistency among copies. Instead of using two-phase commit protocols, lazy replication chooses an optimistic protocol: after the transactions are committed on the source database, the data changes are asynchronously captured, propagated and then applied to the target databases.

The fundamental difference between eager replication and lazy replication is the way to optimize the tradeoff between data consistency and system performance. Whenever increased transaction time is not tolerable (often the case for financial transactions), eager replication is not an option because of the propagation delay incurred over geographic distances. Each 100 km of fiber typically adds about 1 ms of delay [5]. Lazy replication does not impact transaction response time, but introduces two major issues:

(1) Data staleness: After a transaction has committed at the source database, the subsequent data access to the target databases might not return the updated values immediately and consistently. To measure the window of inconsistency between source and target copies, a point-in-time (PIT)-consistency latency is introduced

to describe how much time the target database is behind the source database. We further define consistency group (CG), a set of tables for which transaction consistency is always preserved. The PIT consistency is per CG.

(2) Data loss: An unrecoverable disaster (e.g., earthquake) on the source copy can cause a loss of the data changes that have not been applied to the target databases. PIT consistency latency also largely affects the data loss window in lazy replication. RPO (Recovery Point Objective) in Lazy replication is a non-zero unless there are other means to compensate. Data loss is a function of the replication delay.

To alleviate the impacts of data staleness and data loss, it is highly desirable to reduce the PIT consistency latency especially with the ever-growing data volumes. In the wide area network replication, the PIT consistency latency can reach a non-tolerable value with respect to SLA, during a heavy workload period, particularly batch processes that might update each row of an entire database. To reduce the PIT consistency latency, replication protocols might divide the database objects among several replication channels; potentially have to relax the ACID compliant transaction integrity. At the same time, through synchronization across parallel replication transmission or replay, the eventual data consistency can be ensured at target databases even after disaster recovery (also called 100 % recovery consistency objective). Once all channels have caught up to the same point, consistency is guaranteed. The major benefit is to increase the overall replication throughput through concurrency. Although the negative effects are anomalies (e.g., dirty read) during the replication, most read-only applications can use the data as long as it is not stale beyond a certain threshold, and/or can retry if data has not yet arrived. In these applications, data staleness is more significant than temporary data anomalies.

Our work is applicable to an active-active WAN configuration (where transactions can be executed at either site) presuming that proper transaction routing provides conflict prevention. For discussion simplicity, we present uni-directional replication in an active-query configuration (a.k.a. master-slave [9]) where update transactions are restricted to a designated master copy in one data center and read-only transactions are executed in other data center copies. Upon a failure on the active copy caused by disasters, one of the query copies assumes the master role and takes over the updates.

Figure 1 illustrates a logical architecture of typical parallel lazy data replication between two database systems that potentially reside in two data centers. A parallel replication system can be modeled as a network $G(C \cup A, E)$ with a set of capture $C = \{c_1, c_2, ..., c_s\}$ and a set of apply $A = \{a_1, a_2, ..., a_r\}$. To replicate data changes, a capture agent, such as Capture1 in Fig. 1, captures the committed data changes from the database recovery logs at the source site, packs and sends them over a transport channel. The transport channels manage reliable data transfer between the two sites. An apply agent, such as Apply1 in Fig. 1, applies the changes to the target database. Each capture and apply agent can be attached to a different database node in a cluster. Within the capture and apply agent, whenever possible, multiple threads are used to handle the work with the protection of causal ordering.

The link $(a, c) \in E$ represents a *logical replication channel*, which is an end-to-end replication data path from a log change capture at the source site to a change apply at

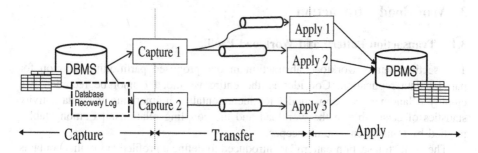

Fig. 1. Logical architecture of parallel lazy replication

the target site. Three channels are shown in Fig. 1. A throughput capacity or bandwidth $BW(a, c)$ measures the maximum data throughput, in bytes/second, of a channel. The value is affected by all the involved components, e.g., source log reader, capture, network, apply, target database, etc. For simplicity, this paper assumes that the effective "bandwidth" is static. All changes within each database object (tables or partitions) are replicated by one channel and this is designated by a preconfigured subscription policy. Each capture agent only captures the changes from its subscribed objects. Transported data changes at target site are subscribed by one or more apply agents on mutually disjoint sets of objects. The entire set of database objects within each replication channel is guaranteed to preserve serial transaction consistency. Hence, the set of database objects $TB = \{tb_1, tb_2, ..., tb_k\}$ that are replicated within the same replication channel (a, c) is called a *consistency group* denoted as $cg(a, c, TB)$.

When a transaction's writeset is split into different consistency groups, the transaction is split into multiple partial transactions with the same source-side commit time. Each partial transaction is replayed at the target as an independent transaction. Replication then operates with eventual consistency: i.e., transaction consistency is guaranteed only when all table changes are replicated up to a common point-in-time. Eventual transaction consistency is suitable for a large number of read-only applications that can use the data as long as it is not stale beyond a certain threshold. Eventual consistency must be restored before write applications can be switched in case of planned site switch or unplanned disaster. Like IBM IIDR Q-Replication [22], replication across consistency groups can be synchronized so that data is not applied to the target DBMS unless it has been received into persistent storage at the target for all consistency groups. Thus, in case of disaster, consistency can be restored by draining the queues for all consistency groups up to a point that is consistent across all queues. When eventual consistency is not acceptable, the target DBMS still can restore the point-in-time consistency using the source-side commit time. Normally, such a consistency recovery mechanism is tightly integrated with the multi-version concurrency control.

In the next two sections, we discuss our solution for optimizing the partitioning of database objects into minimum number of consistency groups to achieve a PIT consistency latency goal based on our workload profiling approach.

3   Workload Abstraction

3.1   Transaction Pattern and Workload Profile

This section details workload abstraction in our proposed partitioning solution for parallel data replication. Considering the entire workload cannot be recorded, we employ a landmark window model for incrementally summarizing the data activity statistics of each table in the workload and the coupling relationships among tables implied by transaction commit scopes.

The term "transaction pattern" is introduced to define a profiled entity that contains statistical and transactional information in a workload. A number of transactions belong to the same transaction pattern if and only if they update exactly the same set of tables, regardless of the specifics such as update sequence, data volume or operation types (insert, update or delete). For instance, given a table set T = {A, B, C, D}, examples of possible transaction patterns are P{A, B, C}, P{A}, P{B, C, D}, P{A, B}, etc. P {A, B, C} and P{A, B} are not considered as the same transaction pattern even though one has a subset of tables of the other. These patterns reflect the hidden table relations recorded in transaction commit scopes. Separate collection of pattern statistics of P {A, B, C} and P{A, B} facilitates the partitioning algorithm in the evaluation of transaction splits. For horizontally partitioned tables, each partition is regarded as an individual physical object. In WPT, the definition of transaction pattern can be easily extended to treat different partitions as different table objects, especially when most of industrial DBMS logs contain the partition identifier. For simplicity, the rest of the paper only discusses tables.

To measure pattern-specific workload size, we record the table-specific statistics, including the numbers of insert, update and delete operations, and data change volumes measured by bytes. For an insert operation, all column values of the new row need to be replicated to the target site. For an update operation, in addition to the values of all the updated columns, the old key values should also be replicated to target sides for row lookup and collision detection. For a delete operation, only the key values need to be replicated. Thus, the workload size depends on the operation types, the actual log contents with the column values, the table definition and the actual column value size. Captured from DBMS catalogs, table schema definitions are used to decode the column values of each row for computing accurate data changes of each operation.

The collected workload patterns and their statistics are modeled by landmark window model. The entire workload is chunked to disjoint pattern snapshots by user-specified time intervals. The collection of all the snapshots constitutes a workload profile.

Figure 2 shows an example transaction pattern entity in a JSON format. It includes information such as snapshot time, transaction pattern identifier (ID), transaction count, the number of tables included, identifiers of the tables, as well as insert, update and delete volumes in bytes. Table-level statistics in a particular transaction pattern include the following content showed in Fig. 2: database ID and table ID; total count of the IUD (Insert, Update, Delete) operations; total bytes of the data that are replicated of the IUD operations; and total bytes of the raw log data in the IUD operations etc.

{ "timepoint":"2014-10-18 16:58:00",
    "tpid":"00000001",
    "tablecount":12,
    "totalcount":500,
    "tablestat":
[{// 1st table statistics.
    "dbid":172,
    "obid":13,
    "isize":0,
    "usize":50000,
    "dsize":10000,
},
{// 2nd table statistics.
    ...
}]}

64 bit

Database id/Tbs id	Table id	Reserved	Reserved
Insert total count			
Update total count			
Delete total count			
Insert total log rate in bytes			
Update total log rate in bytes			
Delete total log rate in bytes			
Insert total raw log rate in bytes			
Update total raw log rate in bytes			
Delete total raw log rate in bytes			

Fig. 2. Example WPT data in JSON format

3.2 Workload Profiling Tool

We implemented a workload profiling tool (WPT) to execute alongside with a commercial database system. The log transaction reader uses the database recovery log APIs to scan through the transaction history log. A transaction pattern control block (TPCB), which stores information of a transaction pattern, is stored in a hash table in an in-memory buffer, as shown in Fig. 3. For every transaction record gathered in log scanning, the statistical information is accumulated if the matching transaction pattern exists in the hash table. If no matching transaction pattern is found, a new transaction pattern entry is added to the hash table. In Fig. 3, TPCB manager maintains a linked list of TPCB buffer. Currently, each buffer has a 64 MB space. When one buffer is full, a new buffer is created and inserted into the list. Using a hash table to store TPCBs ensures that a TPCB entity can be quickly found from these buffers and updated. The address of each TPCB entity is saved in the hash table as a hash value, while the hash key is the hash code of TPCB ID of one TPCB entity. After completing all log records within a snapshot interval, WPT dumps all the gathered information from memory to disk files and then feeds the files to RPA.

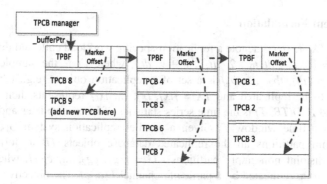

Fig. 3. In-memory TPCB hash table

Since WPT executes in a separate address space from the DBMS, it does not disturb normal database workloads. From a system resource sharing prospective, one can run

WPT at a lower job priority than regular application workloads to avoid or reduce resource contentions. An alternative is to run WPT offline. For example, if the production system disk that contains the database log is mirror-copied to a different system at a local or remote site, WPT can process the log files from the mirrored disk. Multiple WPT instances can also execute concurrently to process log records from different time periods.

WPT and RPA are used in the initial configuration of parallel replication as well as can be applied in the subsequent tuning and optimization process. Figure 4 shows the process of how users can iterate through the capturing, profiling, analysis and database partitioning steps, along with workload growths or new application deployments, to adjust replication groups. RPA supports iterative tuning to help users continuously optimize their partitioning solution. It is up to each replication software whether the redeployment of replication partitioning can be done online or offline.

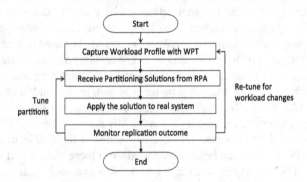

Fig. 4. Iterative parallel replication tuning

4   Replication Partition Advisor Algorithm

4.1   Problem Formulation

Let $WK(TB, TX, T, IUD)$ denote a replication-specific workload collected during a time window $T = \{t_0, t_0 + dt, t_0 + 2 \cdot dt, \ldots t_0 + v \cdot dt\}$, where dt is the sample collection interval; $TB = \{tb_1, tb_2, \ldots, tb_n\}$ is a set of n replication objects (e.g., tables) whose changes are to be replicated and $TX = \{tx_1, tx_2, \ldots, tx_k\}$ represents their transaction activities; and $IUD(TB, T)$ is the time series statistics of inserts, updates and deletes on the tables in a time window T. Given a parallel replication system $G(C \cup A, E)$, RPA-algorithm partitions all the replicated database objects TB to form a set of m mutually disjoint non-empty partitions $CG = \{cg_1, cg_2, \ldots, cg_m\}$, where cg_i is a consistency group replicated by a particular channel $E(a, c)$. The objective is to find a solution such that m is minimal and the worst replication latency in CG is below a user-supplied threshold H.

For a particular replication channel, the PIT consistency latency at a specific time point t_p is the difference between t_p and the source commit time for which all transactions to that point have been applied to the target at time t_p. The latency of each channel is directly related to the logical replication throughput capacity $BW(a, c)$ as well as the size of workload assigned to this channel. The workload size is defined as the number of replicated data bytes. For a specific channel, it can process at most $dt \cdot BW$ bytes within dt seconds. The residual workload will be delayed to the next intervals. Residual workload $RES_{cg,i}$ for a consistency group cg at time $t_0 + i \cdot dt$ is the remaining work accumulated at $t_0 + i \cdot dt$ that has not been consumed by cg_i. Thus, $RES_{cg,i}$ can be computed iteratively by:

$$RES_{cg,i} = max\{(RES_{cg,i-1} + \sum IUD(TB_{cg}, t_0 + (i-1) \cdot dt) - dt \cdot BW), 0\} \quad (1)$$

Assuming data is consumed on a first-in-first-out basis, the PIT consistency latency for cg at time $t_0 + i \cdot dt$ is the time to process the accumulated residue and new activities at $t_0 + i \cdot dt$:

$$PIT_{cg,i} = \left(RES_{cg,i} + \sum IUD(TB_{cg}, t = t_0 + i)\right) / (dt \cdot BW) \quad (2)$$

The maximum PIT consistency latency PIT_{cg} of group cg during the time period is computed as:

$$PIT_{cg} = max\{PIT_{cg,i} | i = 0, 1, 2, \ldots, v\} \quad (3)$$

The maximum PIT consistency latency $PIT_{CG\text{-}max}$ of a set of consistency groups CG is the highest value of PIT_{cg} among all consistency groups in CG.

The objectives of the partitioning optimization can be formulized as follows. Given a workload W, a parallel replication system G and its replication channel bandwidth $BW(a, c)$, and an SLA-driven PIT consistency latency threshold H, the first objective function is defined as:

$$L = \min\{|CG| | \forall CG : PIT_{CG-\max} \leq H\}, \quad (O1)$$

where $|CG|$ is the size of a consistency group set CG, i.e., the number of groups in the set. O1 is to find the partitioning solutions with the lowest number L of consistency groups such that the highest PIT consistency latency of all the replication channels $PIT_{CG\text{-}max}$ is less than or equal to H. Let P_L represent all the partition solutions of group size L and satisfy O1. The second objective is to find a partitioning solution with the minimized number of transaction splits.

$$T\_split = \arg\min_{CG \in P_L} \left\{ \sum_{tx \in TX} \sum_{i=1}^{L} tr^T(cg_i, tx) | tr^T(cg_i, tx) \in \{0, 1\} \right\}, \quad (O2)$$

where $tr^T(cg_i, tx)$ is either 1 or 0 representing whether transaction tx has tables assigned to group cg_i or not. When all the tables in transaction tx are assigned to a single group, $tr^T(cg_i, tx)$ equals 0 for all groups except one. O2 seeks to find the partition solution in P_L such that the aggregated count is minimized. When no transaction split is required, T_split equals the total number of transaction instances in the workload.

4.2 RPA-Algorithm Phase-1: Satisfying PIT Consistency Latency with the Least Groups

Our RPA algorithm consists of two phases: phase-1 is to find a solution that satisfies the first objective O1, and then phase-2 applies a transaction graph refinement approach to achieve the objective O2. The algorithm flow of phase-1 is listed below followed by a description.

RPA-algorithm Phase-1 Steps

1_1.Aggregate the total amount of work $W_{sum} = \sum IUD(TB,T)$ for all tables in $TB = \{tb_1, tb_2, ..., tb_n\}$ and in time period $T = \{t_0, t_0 + dt, t_0 + 2 \cdot dt, ... t_0 + v \cdot dt\}$.

1_2.Compute the lower bound L_{lower} of the number of consistency groups $L_{lower} = W_{sum}/(BW \cdot v \cdot dt + BW \cdot H)$. Set initial CG number $L = L_{lower}$.

1_3.Sort all tables in TB in descending order by each table's peak activity max(IUD) and total activity $\sum (IUD)$, represented by TB_P and TB_T respectively.

1_4.Select a subset TB_{top} consisting of top tables from both list TB_P and list TB_T

1_5.Exhaust all the combinations of placing TB_{top} tables into L groups. Select the placement with the lowest maximum PIT consistency latency and continue to next step.

1_6.Iterate through the rest tables in their descending order in TB_p. Test each table against each consistency group and compute potential maximum PIT latencies $PIT_{pmax\_i}, i = 1,2...L$, for each group. Place the table in the group with the lowest potential PIT_{pmax}. If $PIT_{pmax} > H$ for the selected group, stop and go to 1_8.

1_7.Compute the maximum $PIT_{max\_i}, i = 1,2...L$ for all consistency groups.

1_8.For all consistency groups. If $max(PIT_{max\_i}) > H$ for $i = 1,2...L$, increment $L=L+1$ and repeat steps 1_5 to 1_8 until $max(PIT_{max}) \leq H$. The last L is the minimum number L_{min} of consistency groups.

Given bandwidth $BW(a, c)$ and a user-specified PIT consistency latency threshold H, the first two steps in phase-1 obtain the lower bound L_{lower}, for the number of consistency groups. The lower bound describes the best case scenario: the workload volume distributes uniformly in both table and time dimensions, while the PIT consistency latency reaches the highest at the end of the time window $t_0 + v \cdot dt$ and the residual workload evenly spreads among all channels, i.e. $W_{sum} = L_{lower} \cdot BW \cdot (v \cdot dt + H)$. Starting with this lower bound L_{lower}, the process in steps *1_3* to *1_6* partitions the tables into L_{lower} groups. We then re-examine the actual maximum PIT consistency latency of all groups in steps *1_7* and *1_8*. If the latency is higher than the threshold H, another round of partitioning is performed with the number of groups incremented by 1.

For each fixed group number, the problem becomes to partition n tables into L consistency groups for PIT consistency latency minimization, which is an NP-hard problem [7]. Given that the number of tables in a workload can reach thousands or even more, it is not realistic to exhaust all the partitioning combinations for finding the best among them. Instead, the greedy algorithm is introduced to resolve such a problem [8].

When applying the greedy algorithm, we use a two-step approach for improving the possibility of finding a *global optimal* solution instead of a local optimum. First, using the most active tables TB_{top} (selected in step *1_4*), step *1_5* enumerates all the possibilities of partitioning them into *L non*-empty groups. The number of combinations for such a placement grows rapidly with the numbers of tables and groups. For avoiding an impractically high cost of step *1_5*, the size of TB_{top} is determined based on a reasonable computation time on the system where RPA runs. The best choice from the exhaustive list of placements is the one with the lowest maximum PIT consistency latency. Step *1_5* is then followed by a greedy procedure in step *1_6* that tests each of the rest tables against each consistency group and computes the group's potential new maximum PIT consistency latency contributed by the table. The group with the lowest new maximum PIT consistency latency is the target group for the table placement. The greedy iteration in step *1_6* uses a stronger heuristics for reaching the minimum number of consistency groups, even though it is possible that other partitioning schemes that satisfy objective *O1* (Sect. 4.1) also exist. An added benefit is that this heuristics tends to generate consistency groups with less PIT consistency latency skews among them.

Our approach is particularly effective when there are activity skews among the tables. In fact, such skews are common in real-world applications. Figure 5 shows a customer workload analysis on how tables weight within the workload with respect to total and peak throughputs. A table with a higher *x-axis* value weights more in terms of total throughput than those with lower *x-axis* values. Such a table contributes more to the overall workload volume accumulation and channel saturation. A table with a higher *y-axis* value is more likely to contribute to higher PIT consistency latency at its own peak time. As shown in Fig. 5, tables with higher peak or total throughputs constitute a small fraction in the entire workload. Based on this observation, *step 1_4* selects the top tables with higher total and peak throughputs for enumerative placement tests.

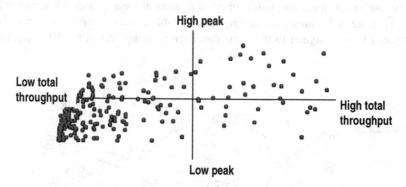

Fig. 5. Table activity distribution in a real-world banking application workload

Throughput-Balancing: An Alternative to PIT Consistency Latency Minimization. Calculation of PIT consistency latencies is impossible when quantified replication bandwidth is unavailable. In this case, the optimization goal of the RPA-algorithm is adjusted to balance the peak volume and total volume given a targeted number of consistency groups. Instead of computing PIT consistency latency, steps *1_5* and *1_6* choose the candidate group based on the accumulated peak volume and total volume after adding a new table. Both factors are positive correlated with the PIT consistency latency. Total throughput-based placement tries to balance utilizations of physical replication channels. Peak throughput-based placement is for capping the highest workload volume among all channels. Understanding workload peaks also facilitates capacity planning and system configuration. This alternative is referred to as the throughput-balancing algorithm (RPA-T-algorithm).

4.3   Transaction Split Reduction

RPA-algorithm phase-1 focuses on reducing PIT consistency latency. This section describes phase-2, which attempts to reduce transaction splits for statistically increasing serial consistency in data replication.

Transaction Graphs. In RPA, we use an undirected weighted graph *TG(TB, TX, T, IUD)* to model tables and their transaction relationships within a workload *WK(TB, TX, T, IUD)*. Each node in the graph represents a table in *TB*. For simplicity, the same notation $TB = \{tb_1, tb_2, ..., tb_n\}$ is also used to represents the graph nodes. The weight of a node is the time series IUD statistics for the table in the workload profile. An edge $e(tb_i, tb_j)$ connecting two nodes tb_i and tb_j denotes that there exists one or more transaction patterns that correlate both tables. The weight of the edge $|e(tb_i, tb_j)|$ is the total transaction instance counts from all the transaction patterns that involve both tables. Figure 6 illustrates an example of a transaction graph with 19 tables. Table *T1*'s weight is associated with a time series statistics *{234, 21, 654, 2556, ..}, which* indicates data activities to be replicated at each time point for *T1*. The weight 731 of edge *e (T1, T2)* means that there are 731 committed transactions involving both *T1* and *T2*.

Because of relational constraints or other reasons, there are cases when transaction consistency must be preserved among certain tables. That means, these correlated tables need to be assigned to the same consistency group. When the RPA-algorithm

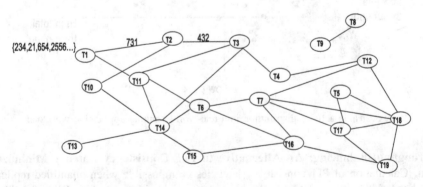

Fig. 6. Transaction graph

builds a transaction graph for a workload, each set of such correlated tables is first merged into a single node with aggregated node statistics and edge weights.

When partitioning a set of table nodes, RPA-algorithm groups the tables to form multiple clusters, which are possibly connected by edges. For a transaction instance that is split into q clusters, the number of edges (of weight 1) connecting these q clusters equals $q \cdot (q-1)/2$. This number monotonically increases with q when $q > 1$. Hence, minimizing the number of split transactions, as formulated by O2 in Sect. 4.1, is equivalent to minimizing the number of edges, or aggregated edge weight. Equivalently, the problem of minimizing transaction splits is a graph-partitioning problem, which is to divide a graph into two or more disconnected new graphs by removing a set of edges. As a classic partitioning problem, *minimum cut graph-partitioning* is to remove a set of edges whose aggregated weight is minimal. A constraint for typical graph-partitioning applications is to balance the total node weight of each partition. Differently, the target of our problem is to minimize the maximum PIT consistency latency among all the groups, each corresponds to an individual consistency group.

General Graph-Partitioning Algorithms. A graph-partitioning problem, as an NP-complete problem in general, is typically solved by heuristics in practice. One widely used algorithm for two-way partitioning (bi-partitioning) is the Kernighan-Lin algorithm (KL algorithm) [13]. It is an iterative improvement algorithm over two existing partitions. It seeks to reduce the total edge cut weight by iteratively swapping nodes in pairs between the two partitions. Fiduccia-Mattheyses algorithm [6] (FM algorithm) further enhances the KL algorithm. By moving a node to a new group, it reduces its edge cut to the other partition while increasing its edge connection to its home partition. It also removes KL algorithm's restriction of moving nodes in pairs. The improved algorithm is referred to as KL-FM algorithm. For large graphs, multi-level bi-partitioning is often applied through graph coarsening and expansion [10]. The quality of their final solutions, which could be a local optimum, is affected by the initial partitioning. Spectral solution [17] can find the global optimum by deriving partitions from the spectrum of the graph's adjacency matrix, but it does not fit our transaction graph model with time series statistics as node weights. Partitioning a graph into more than two partitions can be achieved via a sequence of recursive bi-partitioning. Refinement heuristics for k-way partitioned graph have also been developed [11].

Transaction Split Reduction by Consistency Group Refinement. Before introducing our RPA-algorithm phase-2, we first discuss how to reduce transaction splits between two already partitioned consistency groups by FM algorithms. This process is referred to as an algorithm for 2-CG refinement (CG-RF-2). The process refines the partition via node/table movement. Each move needs to ensure that the PIT consistency latencies for both refined groups remain below PIT_{max} or within a specified margin around PIT_{max}.

Algorithm for 2-way CG refinement(CG-RF-2)

C_1   Create graph representations for each input consistency groups cg_1 and cg_2

C_2   Compute PIT latencies PIT_{cg1} and PIT_{cg2} for cg_1 and cg_2, respectively. Define $PIT_{max}=max(PIT_{cg1}, PIT_{cg2})\cdot(1+\alpha)$ as the upper bound for margin α.

C_3   Compute the gain of each node. The gain for a node table tb_i, as defined in FM algorithm, is computed as the total edge weight between tb_i and all the nodes in the group that tb_i does not belong, subtracted by total edge weight between tb_i and all the nodes in the same group as tb_i, i.e. $g(tb_i)= \sum(|e(tb_i, tb_j)|)- \sum(|e(tb_i, tb_k)|)$ where tb_j belongs in the different group than tb_i, and tb_k belongs in the same group as tb_i. The intuition is that if $g(tb_i)$ is positive, moving tb_i from its current group to the other group reduces the edge cut between the two groups.

C_4   Find the node n_1 with the maximum gain g_1 and whose move from its current group to the other allows each group's PIT consistency latency remains below the PIT_{max} value from C_2. Lock node n_1, mark its movement from its current group to the other as an element mv_1 and store in the moving list mv_list. In some cases, the gain of node n_1 is non-positive. However, it is still moved with the expectation that the move will allow the algorithm to "escape out of a local minimum".

C_5   Update the gains of all the nodes that are connected to n_1 due to its movement.

C_6   Repeat C_4 and C_5 for the rest of the nodes until all the nodes are locked. All movements are stored in $mv\_list\{mv_1,...,mv_n\}$ in the order that they are found. The gains corresponding to these node moving steps is $\{g_1,g_2,...,g_n\}$.

C_7   Find the best sequence of $mv_1, mv_2,...,mv_k$ $(1\leq k\leq n)$ such that $\sum(\{g_1,g_2,...,g_k\})$ is maximum and positive.

C_8   Mark the move of these k tables permanent. The refined groups are cg_1' & cg_2'.

C_9   Free all the locked nodes.

C_10 Repeat steps C_3 to C_9 until no move can be found in C_7.

The PIT consistency latency upper bound in *C_2* is set to preserve the optimization objective and speed up the algorithm convergence. When the two input groups are produced by RPA-algorithm phase-1 and α is set to 0, CG-RF-2 algorithm preserves the same maximum PIT consistency latency value from phase-1 while refining the groups for transaction split minimization. When $\alpha > 0$, the PIT consistency latency constraint is relaxed and potentially more nodes are moved to reduce transaction split. Alternatively, a user-supplied PIT threshold H can be used as the constraint.

In some cases, the two-step procedure of bi-partitioning and refinement can be used recursively to create a higher number of partitions, given that the refinement constraint can be distributed along the recursion paths. Such an approach works for throughput-balancing partitioning optimization, i.e. the alternative algorithm RPA-T. However, PIT consistency latency is not a constraint measure that can be easily distributed while still guaranteeing convergence during recursive bi-partitioning. Therefore a non-recursive approach is needed.

RPA-Algorithm Phase-2: K-Way Consistency Group Refinement for Transaction Split Reduction. This section presents the phase-2 of our RPA-algorithm for transaction split reduction. The algorithm (called CG-RF-k) is derived from the k-way refinement algorithm proposed by Karypis et al. [11].

RPA-algorithm Phase-2 (CG-RF-k)

Ck_1 For the k consistency groups $cg_1, ..., cg_k$ created by RPA-algorithm phase-1, create the graph representation for the workload and these k partitions.

Ck_2 Iteration through all the nodes, find the set N_e of all the nodes that each has edge connections to other groups that it does not belong to. Compute the gain for each element in N_e, denote a gain as g(tb_i, cg_m) in which cg_m is a group that node (table) tb_i does not belong but has edge connections to one or more of its nodes. The gain is computed the same as algorithm CG-RF-2 step C_3.

Ck_3 Compose subset N_e' of N_e with nodes that only have positive gains.

Ck_4 For each node tb_i in N_e', test it with its connected groups for potential new PIT latencies. Among those groups whose potential new PIT latencies are below the user specified threshold H, select the group with the largest positive gain for tb_i to move into. If none of the group qualifies the PIT threshold requirement, do not move tb_i.

Ck_5 Update the gains of all the affected nodes due to the move of tb_i, including tb_i. Updates N_e' following the same criteria as in Ck_3.

Ck_6 Repeat steps Ck_4 and Ck_5 until there is no node in N_e'.

RPA-algorithm phase-2 starts its refinement process from the partitioning result of phase-1, which finds the minimum number of groups while satisfying maximum PIT consistency latency threshold. Every node move seeks to reduce the positive gains, i.e. trading higher inter-group edge cut weight with lower intra-group edge cut weight. This process keeps reducing the transaction split count until reaching the lowest.

4.4 Real-Time Partitioning Evolution

In a production environment, partitioning can evolve to adapt to the changing workload patterns and system environment. Both workloads and underlying resources are self-governing agents that are autonomous from replication software. The maintenance issues are even more important than the initial construction, especially in such a dynamic environment. When the detected changes lead to the real time PIT consistency latency increase above a specific threshold H, re-partitioning is executed with the following three-step tasks:

(a) Re-computing the partitioning of database objects with the latest data.
(b) Draining the on-going replication. Since the queued work can be congested at anywhere in between the capture and apply, in reality it is very complicated to re-direct those to different replication channels. During draining, capturing of new workload from current channels is suspended to allow the existing channels to finish replicating queued workload. Newly added channels can start as soon as possible since there is no queued workload in those channels.
(c) Deploying the new replication partitioning configuration into use.

The first two steps can be applied in parallel. Since step (a) usually consumes a considerably shorter time than step (b), we only concern the PIT impact by step (b) in the following discussion. Denote the period using the earlier consistency group configuration P_1, and the period with the modified consistency group configuration P_2. Computing residue throughput and PIT in the situation of re-partitioning is a bit different from what is described in Sect. 4.1. Denoting the PIT consistency latency of the old-period consistency group in the last period as PIT_{p1}, the time for draining the old-period workload td is computed as:

$$td = PIT_{p1} \cdot dt \tag{4}$$

The residual workload REScg,i in the draining period and later should be computed iteratively according to the value of time interval index i:

$$RES_{cg,i} = \begin{cases} RES_{cg,i-1} + \sum IUD(TB_{cg}, t_0 + (i-1) \cdot dt), & i \le PIT_{p1} \\ \max\{(RES_{cg,i-1} + \sum IUD(TB_{cg}, t_0 + (i-1) \cdot dt) - dt \cdot BW), 0\}, & i > PIT_{p1} \end{cases} \tag{5}$$

Accordingly, $PITcg\text{-}i$ computation should be adjusted as follows:

$$PIT_{cg,i} = \begin{cases} PIT_{cg-old} + (RES_{cg,i-1} + \sum IUD(TB_{cg}, t = t_0 + i))/(dt \cdot BW), & i \le PIT_{p1} \\ (RES_{cg,i-1} + \sum IUD(TB_{cg}, t = t_0 + i))/(dt \cdot BW), & i > PIT_{p1} \end{cases}$$
$$\tag{6}$$

5 Experiments and Analysis

We applied our work to a batch workload and an OLTP workload. The batch workload is from a banking business and we collected the WPT data from an offloaded production DBMS recovery log. For the OLTP workload, we expanded the schema of TPC-E benchmark [24] and simulated workload profile data for analysis. In both experiments, the analysis processes complete within minutes.

5.1 Transaction Split Avoidance Algorithm

For the purpose of comparison and establishing experimental baselines, we devised an algorithm named Transaction Split Avoidance (TSA). This algorithm assists studying the trade-offs between transaction split and either replication latency or throughput-balancing in all our experiments. We implemented TSA algorithm which seeks reducing PIT consistency latency under the constraints that no transaction split is allowed. Using a transaction graph, the TSA algorithm groups nodes (database objects) connecting to each other directly or indirectly into one *virtual table* by breadth-first search, and then applies the PIT minimization algorithm to these virtual tables.

T_1 *Create a graph representation including all tables as nodes and their transaction-related connections as edges.*

T_2 *Traverse all nodes (tables) in the graph, and partition them into a set of connected sub-graphs with no connection between any two sub-graphs.*

   T_2-1 *Starting from any table node tb_i in the* graph, *perform depth-first-traversal to find all table nodes that are reachable directly or indirectly from node tb_i via connecting transaction edges and other connected nodes.*

T_3 *Merge tables in each sub-graph into one virtual table by accumulating their workloads. VT (see formula 7) is a collection of such virtual tables. The time series statistics on a virtual table in the time window T is an accumulation (see formula 8) from all real tables that constructs the virtual table.*

$$VT = \{vt_1, vt_2, ..., vt_l\} \tag{7}$$

$$\begin{cases} IUD(vt_j, T) = \sum IUD(tb_k, T), \\ \quad tb_k \in vt_j, \\ T = \{t_0, t_0 + dt, t_0 + 2 \cdot dt, ..., t_0 + v \cdot dt\} \end{cases} \tag{8}$$

T_4 *Sort all virtual tables in descending order by each virtual table's peak PIT consistency latency max(PIT), represented by VT_{pit}.*

T_5 *Select a subset VT_{top} consisting of top virtual tables from list VT_{pit}.*

T_6 *For a specified number of consistency groups L, exhaust all the possible placement of grouping VT_{top} into L CGs. Select the placement with the lowest maximum PIT consistency latency and assign real tables associated with those virtual tables into corresponding consistency groups.*

T_7 *Iterate through the remaining virtual tables in their descending order in VT_{pit}. Test each virtual table against each consistency group and compute potential maximum PIT latencies $PIT_{max\_i}, i=1,2,.....,L$, for each group. If there are some groups whose PIT latencies are not impacted by adding this virtual table, place tables belonged to this virtual table into the group with lowest PIT consistency latency. Otherwise, place these tables into the group whose corresponding PIT_{max} is the lowest.*

5.2 Experiment with a Large Bank Batch Workload

This workload profile was collected from a database log representing a four-hour batch processing window with 1 min sample interval. There are 824 tables with active statistics among a total of 2414 tables, and 5529 transaction patterns are discovered from 12.7 million transaction instances. The number of tables correlated by transaction patterns varies between 1 and 27 within the histogram shown in Fig. 7.

We apply the RPA-algorithm with a replication bandwidth $BW = 5$ MB/s. To put in prospective, this bandwidth is equivalent to insert 50 K 100-byte records per second into a database. Starting from the lower bound of 3 consistency groups following step 1_2 of RPA-algorithm phase-1, Fig. 8 shows the maximum PIT consistency latency of each group, in the unit of a sample interval, when the workload is partitioned into 4, 6, 8, 10 or 12 groups. As the number of consistency groups increases, the PIT latencies are reduced for each configuration. The reason that the three highest PIT consistency latency values remain unchanged in 8-, 10- and 12-group cases is because these three groups are assigned with only one volume-heavy table to each group. To further reduce point-in-time latency, single channel replication bandwidth has to be increased by improving the underline replication technologies in network, database, and replication software.

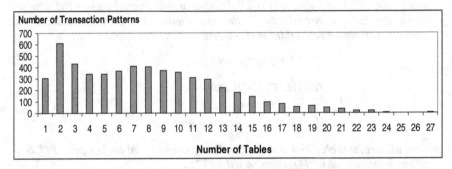

Fig. 7. Distribution of transaction patterns over the number of tables in a batch workload

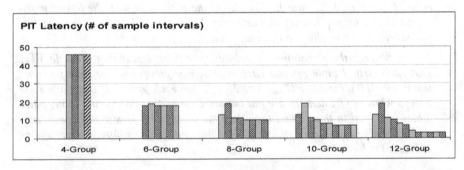

Fig. 8. Partitioning result of batch workload with RPA-algorithm phase-1

Next we apply both phase-1 and phase-2 of RPA-algorithm to reduce transaction splits for a given PIT consistency latency threshold $H = 60$ (1 h). The lowest number of consistency group for this threshold is four from phase-1. Figure 9 shows the result of phase-2. The first chart in Fig. 9 shows the maximum PIT consistency latency of each consistency group using different variations of RPA-algorithm such as phase-1 only, phase-1 plus phase-2 with allowed increase in PIT consistency latency within 0 %, 10 % and 20 % margin, as labeled accordingly in the chart. The second chart in Fig. 9 shows the transaction split distribution in terms of number of groups. Note that splitting into one group means no splitting. TSA algorithm's results are also provided for comparison.

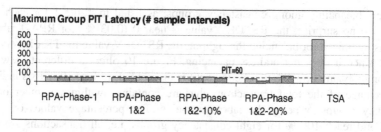

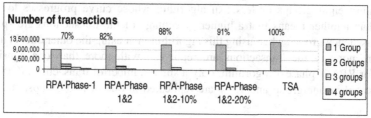

Fig. 9. Partition and transaction split results with RPA-algorithm phase-1 & phase-2 (4 CGs)

The charts show that when phase-2 is used after phase-1, the percentage of non-splitting transactions increases from 70 % with "RPA_Phase1" to 82 %, 88 % and 91 % respectively for RPA_Phase1&2, RPA_Phase1&2-10 % and RPA_Phase1&2-20 %. With the TSA algorithm, all the transactions are non-splitting; however the maximum PIT consistency latency reaches unacceptably high of over 450 1-min sample intervals. In addition to demonstrating that RPA-algorithm can effectively reduce transaction split, the result provides trade-offs study between transaction split and PIT consistency latency.

5.3   Experiment with an OLTP Workload

TPC-E is a newer OLTP data centric benchmark. Its processing is composed of both READ-ONLY and READ-WRITE transactions. Only the READ-WRITE transactions with data changes are used in our study. The TPC-E table schema consists of 33 tables, and 23 of which are actively updated during the transaction execution flows.

To simulate more complex real-world workloads, we expanded the schema by increasing the number of tables by 30× as well as increasing transaction correlations among the tables. Based on the augmented schema and workloads, as well as TPC-E specification on how the tables are updated, we generated a simulated workload profile data with 155 transaction patterns and over 6 million transactions.

OLTP workloads usually update the smaller amount of data within the scope of a committed transaction. Since the volume is lower than the batch, we experiment with our alternative throughput-balancing algorithm (RPA-T-algorithm) and to partition the tables and balance total throughput among 8 consistency groups.

The analyses of the partitioning results using RPA-T phase-1 and RPA-T phase-1&2 are shown in Table 1 and Fig. 10. To be more intuitive, relative standard deviation (RSTDEV = standard deviation/mean) is used to evaluate the effectiveness of throughput-balancing among consistency groups, as listed in Table 1 for each algorithm. With no surprise, the RSTDEV value is near 0 (0.03 %) for RPA-T phase-1 since it is optimized for balancing throughput; the RSTDEV value for TSA is very high (282 %) since it does not address balancing. Figure 10 offers a different view than Fig. 9 for analyzing how the transaction split is distributed. In Fig. 10, y-axis indicates the percentage of the total transactions that are contained within x number or less consistency groups, x being the label on x-axis. The percentage values on y-axis increase and reach 100 % for eight consistency groups, i.e. all transactions are replicated within eight groups or less. An algorithm whose curve progresses to 100 % slower than another means that a higher percentage of the transactions are split into more consistency groups when using this algorithm than using the other one. With TSA algorithm, none of the transactions are replicated with more than one consistency group. For RPA-T phase-1 algorithm, only a small number of transactions (0.0015 %) are replicated in one group and 15 % are replicated in one or two groups, etc.

Table 1. Throughput RSTDEV for different algorithm

	RPA-T	TSA	RPA-T phase-1&2 (throughput trade-off %)		
	phase-1		0 %	1 %	5 %
RSTDEV	0.03 %	282 %	0.03 %	1.15 %	7.84 %

Like RPA-algorithm, RPA-T phase-2 seeks to reduce transaction split count among consistency groups generated by RPA-T phase-1. Table 1 and Fig. 10 show that the RPA-T phase-1&2 (0 %) curve progresses only marginally faster than RPA phase-1. Because the activities in this workload are uniformly distributed among different tables and along the time dimension, by not allowing throughput trade-offs (0 %), it limits the number of tables that can be moved during refinement. For further transaction split reduction, more trade-offs are needed on throughput-balancing constraint. As observed from Fig. 10, with 1 % and 5 % allowed adjustment on throughputs constraint during each refinement step, there are significant increases in the number of transactions that

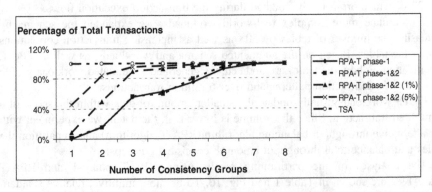

Fig. 10. Transaction split result for OLTP workload

are replicated using less consistency groups. For example, 49.2 % and 84.0 % of transactions are replicated with two consistency groups or less, respectively using RPA-T phase-1&2 (1 %) and RPA-T phase-1&2 (5 %). The trade-offs increase the throughput deviations among groups, e.g. to RSTDEV = 1.15 % for RPA-T phase-1&2 (1 %) and RSTDEV = 7.84 % for RPA-T phase-1&2 (5 %). Such deviation is less significant compared to the reduction in transaction splits.

5.4 Simulation of Partitioning Evolution

We conducted simulation to demonstrate how our RPA tool is applied when real-time workload fluctuates and deviates from the previous profile. In the experiment, we devise a monitor to check the PIT consistency latency of all consistency groups periodically using timestamp information associated with the workloads. When a consistency group's PIT consistency latency is identified higher than the threshold due to the change of run-time environment, re-partitioning is triggered. In this experiment shown in Fig. 11, four consistency groups were used initially in capture and apply pairs (C_1, A_1), (C_2, A_2), (C_3, A_3), and (C_4, A_4), and each replicated a set of database objects not overlapping with the other groups. At the very beginning, the bandwidth of each channel was 300 KB/s. As we can see from Fig. 11, the maximum PIT consistency latency among consistency groups was under the threshold 9 during the first 60 time units (In this experiment, 1 time unit = 1 min). However, at time 60, the system replication bandwidth decreased to 180 KB/s and caused the PIT latencies of all consistency groups to increase significantly. As a result, the PIT consistency latency of one consistency group exceeded the threshold at time 62 and triggered the re-partitioning. We re-applied RPA with the changed bandwidth value and adjusted the partitioning scheme to use as many as eight consistency groups. At the same time, draining started for the queued workload that had been captured but not applied from (C_1, A_1), (C_2, A_2), (C_3, A_3), and (C_4, A_4). At time 69, the draining of all database objects

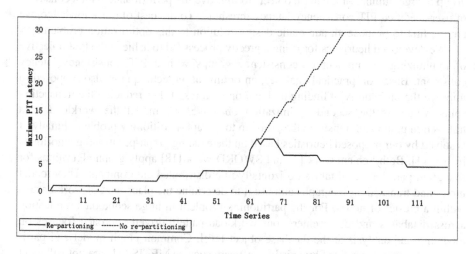

Fig. 11. Simulation of re-partitioning

was finished. All eight replication channels started working to capture and apply any changes since timestamp 62 in RDBMS transaction log. As mentioned in Sect. 4.4, the newly added four channel could have started earlier but for simplicity we started all after the last draining was finished. With the appliance of new configuration, the *PIT consistency latency* had dropped below the threshold after time 69. On the other hand, Fig. 11 also shows that the maximum PIT consistency latency would potentially be beyond 30 if no adjustment was made.

6 Related Work

Database replication is a key technology and a challenging problem for achieving data serving high availability and disaster tolerance [9, 12]. Prior works attempt to address various aspects of replication such as transaction consistency protocols, scalability and performance, etc. (e.g. [14, 15, 19]). In "share nothing" architecture, data replication is used to move data elements among processing nodes to mitigate system failure or to localize transaction processing for better performance [1]. In Spanner [3], synchronous replication is used to achieve transaction consistency in globally distributed data stores. The work in Schism [4] proposes an approach of workload-driven, graph-based replication and partitioning combined with explanation and validation. The work in SWORD [18], targeting data-as-a-service in a cloud environment, achieves higher scalability over prior work with a set of new techniques and introduces incremental re-partitioning. Both works build replication components within the data-serving software.

As reported by Cecchet et al. [1], various challenges still exist when applying database replication in commercial business environments. Motivated by a real-world problem, this paper aims at optimizing middleware-based parallel data replication, especially in a long-distance multi-data-center setting. By filling a gap in understanding database objects affinities with transaction workloads, our work investigates how to group a large number of database objects to improve the performance with a constraint of user-specified PIT consistency latency threshold. To the best of our knowledge, we are the first to propose an automatic design solution to this optimization problem.

We developed heuristics for using a greedy process [8] to achieve the first objective of minimizing the number of consistency groups with a PIT consistency latency constraint. Based on practical analyses, an optimization technique is also proposed to improve the probability of finding a global optimal result. For reducing the transaction splits, which is the second optimization objective, we model the workload as a transaction graph and transform the problem to a graph-partitioning problem. Finally, it is solved by our proposed heuristics based on the existing graph-partitioning algorithms [6, 13, 11]. Both Schism work [4] and SWORD work [18] apply graph algorithms for fine-grain partitioning of tables horizontally in a distributed environment. They model tuples and transactions as graphs and use it to determine the placement of work or data within a cluster of nodes. For the partitioning problem in large-scope data replication across databases and data centers, our workload-pattern-driven approach focuses on modeling and analysis at the database object level. Common graph model and partitioning algorithms provided by existing software such as METIS [21] are not sufficient for our problem. This is because, in order to address workload fluctuation and address PIT objective, we need to model a workload transaction graph using time series statistics from the tables and the transactions and the computation of PIT consistency latency is iterative with respect to workload volume and time. Our algorithm also needs to introduce problem-related heuristics during the partitioning phase to handle multiple optimization objectives and trade-offs under PIT consistency constraint.

7   Conclusion and Future Work

Large-scale database replication is essential for achieving IT continuous availability. This paper presents a workload discovery and database replication partitioning approach to facilitate parallel inter-data-center data replication that is applicable to both share-nothing and share-disk databases. Our design and algorithms are demonstrated with a real customer batch workload and a simulated OLTP workload. In practice, the work has been applied to a real-world business applications environment. For future work, we plan to further fine-tune the optimization model for the replication stack.

Acknowledgements. We would like to thank Austin D'Costa and James Z. Teng for their insights.

References

1. Cecchet, E., Candea, G., Ailamaki, A.: Middleware-based database replication: the gaps between theory and practice. In: SIGMOD (2008)
2. Codd, E.F.: The Relational Model for Database Management, Version 2. Addison-Wesley, New York (1990). ISBN: 9780201141924
3. Corbett, J.C., et al.: Spanner: Google's globally-distributed database. In: OSDI (2012)
4. Curino, C., Jones, E., Zhang, Y., Madden, S.: Schism: a workload-driven approach to database replication and partitioning. Proc. VLDB **3**, 48–57 (2010)
5. DeCusatis, C.: Handbook of Fiber Optic Data Communication: A Practical Guide to Optical Networking, 4th edn. Academic Press, London (2013). ISBN: 10 0124016731
6. Fiduccia, C.M., Mattheyses, R.M.: A linear-time heuristic for improving network partitions. In: Proceedings of the 19th Design Automation Conference, pp. 175–181, January 1982
7. Garey, M.R., Johnson, D.S.: Computers and Intractability; A Guide to the Theory of NP-Completeness. W. H. Freeman & Co., New York (1990)
8. Graham, R.L.: Bounds on multiprocessing anomalies and related packing algorithms. In: AFIPS Spring Joint Computing Conference, pp. 205–217 (1972)
9. Gray, J., Helland, P., O'Neil, P., Shasha, D.: The dangers of replication and a solution. In: SIGMOD (1996)
10. Karypis, G., Kumar, V.: A fast and high quality multilevel scheme for partitioning irregular graphs. SIAM J. Sci. Comput. **20**(1), 359–392 (1998)
11. Karypis, G., Kumar, V.: Multilevel algorithms for multi-constraint graph partitioning. In: Proceedings of the 1998 ACM/IEEE Conference on Supercomputing (1998)
12. Kemme, B., Jiménez-Peris, R., Patiño-Martínez, M.: Database replication. Synth. Lect. Data Manag. **5**, 1–153 (2010). Morgan & Claypool Publishers

13. Kernighan, B.W., Lin, S.: An efficient heuristic procedure for partitioning graphs. Bell Syst. Techn. J. **49**, 291–307 (1970)
14. Lin, Y., Kemme, B., Patiño-Martínez, M., Jiménez-Peris, R.: Middleware based data replication providing snapshot isolation. In: SIGMOD (2005)
15. Patiño-Martínez, M., Jiménez-Peris, R., Kemme, B., Alonso, G.: MIDDLE-R: consistent database replication at the middleware level. ACM TOCS **23**(4), 375–423 (2005)
16. Pavlo, A., Curino, C., Zdonik, S.B.: Skew-aware automatic database partitioning in shared-nothing, parallel OLTP systems. In: SIGMOD (2012)
17. Pothen, A., Simon, H.D., Liou, K.: Partitioning sparse matrices with eigenvectors of graphs. SIAM J. Matrix Anal. Appl. **11**(3), 430–452 (1990)
18. Quamar, A., Kumar, K.A., Deshpande, A.: SWORD: scalable workload-aware data placement for transactional workloads. In: EDBT (2013)
19. Serrano, D., Patiño-Martínez, M., Jiménez-Peris, R., Kemme, B.: Boosting database replication scalability through partial replication and 1-copy-snapshot-isolation. In: Proceedings of the 13th PRDC (2007)
20. Stonebraker, M.: The Case for Shared Nothing. IEEE Database Eng. Bull. **9**(1), 4–9 (1986)
21. http://glaros.dtc.umn.edu/gkhome/views/metis
22. IBM Infosphere Data Replication. http://www-03.ibm.com/software/
23. Oracle GoldenGate. http://www.oracle.com/technetwork/middleware/goldengate/
24. http://www.tpc.org/tpce/

Anonymization of Data Sets with NULL Values

Margareta Ciglic, Johann Eder$^{(\boxtimes)}$, and Christian Koncilia

Department of Informatics Systems, Alpen-Adria-Universität Klagenfurt,
Klagenfurt, Austria
{margareta.ciglic,johann.eder,christian.koncilia}@aau.at

Abstract. Releasing, publishing or transferring microdata is restricted by the necessity to protect the privacy of data owners. k-anonymity is one of the most widespread concepts for anonymizing microdata but it does not explicitly cover NULL values which are nevertheless frequently found in microdata. We study the problem of NULL values (missing values, non-applicable attributes, etc.) for anonymization in detail, present a set of new definitions for k-anonymity explicitly considering NULL values and analyze which definition protects from which attacks. We show that an adequate treatment of missing values in microdata can be easily achieved by an extension of generalization algorithms. In particular, we show how the proposed treatment of NULL values was incorporated in the anonymization tool ANON, which implements generalization and tuple suppression with an application specific definition of information loss. With a series of experiments we show that NULL aware generalization algorithms have less information loss than standard algorithms.

Keywords: Privacy · k-anonymity · NULL values · Missing values

1 Introduction

Detailed data collections are an important resource for research, for fact based governance, or for knowledge based decision making. In the field of statistical databases any collection of data with detailed information on entities, in particular persons and organizations, is called microdata.

A crucial requirement for the release of microdata is the preservation of the privacy of the data owners, which is protected by laws and regulations. Furthermore, for data collections requiring the willingness of data owners to share (donate) their data, studies [9] clearly indicate that the protection of privacy is one of the major concerns of data owners and decisive for a consent to donate data [13]. For protecting privacy from linkage attacks the concept of k-anonymity [25] received probably the widest attention. Its core idea is to preserve privacy

The work was supported by *Austrian Ministry of Science and Research* within the project BBMRI.AT (GZ 10.470/0016-II/3/2013) and *Technologie- und Methodenplattform für die vernetzte medizinische Forschung e.V.* (*TMF*) within the project ANON.

© Springer-Verlag Berlin Heidelberg 2016
A. Hameurlain et al. (Eds.): TLDKS XXIV, LNCS 9510, pp. 193–220, 2016.
DOI: 10.1007/978-3-662-49214-7_7

by hiding each individual in a crowd of at least k members. Many anonymization algorithms implementing these concepts were developed.

Surprisingly, neither the original definition of k-anonymity nor any of the many anonymization algorithms deals explicitly with unknown, or missing values (NULL values in database terms) in microdata. We could not find a single source discussing the problem of NULL values in microdata for anonymization. Recent surveys [19] or textbooks [12] do not mention NULL values or missing values. However, all techniques and algorithms we found, explicitly or implicitly require that all records with at least one NULL value have to be removed from a table before it can be anonymized ([2, 15–18, 23, 29, 32], and many more). There is only some treatment of NULL values in form of suppressed values, i.e. NULL values resulting from removing ("suppressing") data in the course of anonymization procedures. Attacks on suppressed rows can be found in [22, 30]. [1, 7, 21] discuss suppression of values in single cells. However, neither of these approaches discusses the problem of missing values in the original data or of non-existing values due to non-applicable attributes.

NULL values, nevertheless, are not exceptional in microdata, e.g. they appear frequently in data sets for medical research [8, 10, 28, 33]: Some attributes might not be applicable for each patient. A patient might have refused to answer some questions in a questionnaire or could not be asked due to physical or mental conditions. In an emergency some test might not have been performed, etc.

Anonymization by generalization and suppression of data cause loss of information. The aim of reducing this information loss triggered many research efforts. The ignorance of NULL values in anonymization algorithms results in dropping rows from a table, causing a considerable loss of information. Furthermore, dropping rows with NULL values also could introduce some bias in the data set, which is not contained in the original table. This is of course unfortunate for further analysis of the data (for example in evidence based medicine) and might compromise the statistical validity of the results (e.g. dropping rows with a NULL value in the field *occupation* would skip all children from the data set and introduce an age bias, which was not present in the original data set).

This paper is an extension of [4]. We provide a thorough grounding for the treatment of NULL values in anonymization algorithms. We show that we can reduce the problem of NULL values in k-anonymity to different definitions of matching between rows of a data set based on extending the comparison of values and NULL values. We show that generalization algorithms, widely used for anonymization, can be easily extended to cover NULL values and we show that this extension reduces information loss during anonymization significantly. In particular, we show how the treatment of NULL values was incorporated into the anonymization tool ANON, which implements a generalization algorithm with tuple suppression, which optimizes application specific usability of the anonymized data by minimizing the information loss, which is defined by the user, specifically for the application and the intended use of data [3, 27, 28].

k-anonymity (which is defined on the quasi identifiers) has to be complemented with ℓ-diversity for sensitive attributes [18] to avoid that an adversary might infer data of individual. In this paper, however, we focus on *k*-anonymity and only briefly treat ℓ-diversity, as it is implemented in ANON.

2   *k*-Anonymity Revisited

A detailed collection and representation of data on information subjects is called microdata - as opposed to data in less detail like statistical data. For this paper a microdata table is a multiset of rows [14]. We can classify the attributes in the schema of such table in four categories: (1) identifiers: all attributes which uniquely identify a row in the table, (2) quasi identifier: all attributes which an adversary might know, (3) sensitive attributes: attributes with values that should not be inferable by an adversary and (4) all other data. For this paper we assume that the identifiers have already been removed from a table and that the schema of a table includes a set of quasi identifiers Q, which we denote by $Q_1, \ldots, Q_n$.

The aim of anonymization is to assure that a table can be published without opening an adversary the possibilities to gain additional knowledge about the data subjects.

Table 1 shows our running example for such a table with the quasi identifiers `Gender`, `Height`, `Job`, and `ZIP` and the sensitive attribute `Condition`.

Table 1. Original table

	Gender	Height	Job	ZIP	Condition
A	f	165	NULL	9020	Cancer
B	m	187	Mayor	9020	Hepatitis
C	f	163	Clerk	9020	Flu
D	m	NULL	Technician	9020	Pneumonia
E	m	183	NULL	9020	Malaria
F	m	189	Pilot	9020	Gastritis

Samarati and Sweeney [25] proposed an approach to preserve the privacy of a data owner by hiding each data owner in a crowd of at least k individuals, such that an adversary might not get detailed information about an individual, but only information about a group of k individuals. The larger the k, the smaller the possible information gain of an adversary.

In [26] *k*-anonymity is defined as follows: 'Each release of data must be such that every combination of values of quasi identifiers can be indistinctly matched to at least k individuals'. The term *indistinct match* is not defined explicitly, nevertheless, it is clear from the context that two rows match, if they have identical values in the quasi identifiers. However, missing values are not mentioned. We basically follow this definition here, and analyze, how rows of a table match in case some values are NULL. Hence we formalize the notion of *k*-anonymity, dependent on some match operator.

Definition 1 *(k-Anonymity).* *Let T be a table and Q the set of quasi identifier attributes and let $\sim$ be a match predicate on T. T is k-anonymous with respect to $\sim$, iff $\forall t \in T : |\{t'|t \sim t'\}| \geq k$.*

k-Anonymity as well as ℓ-diversity can be achieved with two basic techniques, *generalization* and *suppression*. Both techniques decrease information content of the data to meet the required degree of privacy. They can be used separate or in combination. Generalization [26,30] replaces the values of quasi identifiers (QID) with more general values defined in the generalization hierarchies (taxonomy trees or intervals with step definitions) for all QIDs. The leaves of a generalization hierarchies are the original values of the domain of an attribute, the top level is a single value ALL, which has no information content. The generalization hierarchy and its corresponding domains of the QID *Job* of the running example (Table 1) are shown in Fig. 1.

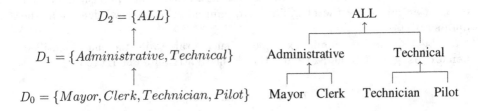

Fig. 1. Generalization hierarchy and its corresponding domains (generalization levels) of the QID *Job* of the running example

The anonymization procedure in general is as follows: When a row does not match at least $k - 1$ other rows, then some attribute values are generalized, i.e. replaced with the parent of this value in the generalization hierarchy defined for each domain (resp. each attribute). In the case of local recoding this is done for individual rows, for global recoding or full-domain generalization scheme the generalization is performed for all the rows [12,31]. This is repeated until the table is k-anonymous or the highest level of generalization is reached in all attributes. A shortcoming of this method is that outlier tuples in the microdata can lead to a very coarse grain generalization. Outlier tuples are those, which hardly match any other tuples. If they remain in the table that is being anonymized, the overall information loss increases. To avoid information loss caused by such outliers, full domain generalization is mostly accompanied with row suppression, where given fraction of rows might be suppressed, i.e. these rows are removed from the table or all their values are replaced by ALL (resp. NULL). To avoid attacks at least k rows have to be suppressed [30].

It is easy to see that in general several tables qualify as result. The aim is now to compute the table with the lowest information loss. The problem is known to be NP complete [21]. However, for global recoding the complexity of the method is exponential in the number of quasi identifiers and their generalization hierarchy height and not in the number of tuples. Many algorithms have been proposed, which apply heuristics to reduce complexity and which apply different measures for information loss to efficiently compute "good" anonymized tables.

In contrast to generalization, suppression does not transform the values to other, more general values, but simply deletes (eliminates) them. Suppression can be undertaken on single values (called cell suppression), on whole tuples (called tuple suppression) or on whole attributes (called attribute suppression). The impact of the attribute suppression is the same as the one of the generalization of an attribute to the top level. Approximation algorithms that use cell suppression are described in [1, 21]. In combination with generalization [26, 30], tuple suppression can be used to eliminate outlier tuples, while the remaining tuples get generalized.

3 *k*-Anonymity with NULL values

3.1 NULL Values

NULL values [20] are the standard way of representing missing information in database tables. We can distinguish three kinds of NULL values: (1) attribute not applicable: in this case there is no value for this attribute for this row in the world represented in the database. (2) missing value: there exists a value in the world, but it is not contained in the database. (3) no information: it is not known whether the value exists in the world or not. In SQL the semantics of NULL is "no information".

For the following considerations we follow the treatment of NULL values in SQL [14]. This means in particular, that a comparison of a NULL value with any other value never results in TRUE and there is a special unary predicate *is* NULL to test for NULL values.

3.2 Matching NULL Values

For matching of NULL values we have the following options:

- basic match: NULL values do not match with NULL values, nor with any other value.
- extended match: NULL values match with NULL values.
- maybe match: NULL values match with any value including NULL values.

In the original definition of *k*-anonymity and in the current anonymization algorithms basic match is used. It is in accordance with the definition in SQL, where 'A = B' is not true, if A or B are NULL values. Extended match treats a NULL value like any other value. Maybe match sees NULL values as wildcards for matching. It corresponds to Codd's maybe selection [6], where rows are returned, if the selection predicate is true for a substitution of the NULL values.

3.3   Basic Match

We call the match used in [26], where rows with NULL values are discarded, *basic match*, and formally define it as follows:

Definition 2 *(Basic match)*. *Let T be a table and Q the set of quasi identifier attributes.*

$$t_1 \sim_b t_2 :\Longleftrightarrow \forall q \in Q : t_1[q] = t_2[q]$$

For illustrating the different match definitions and their consequences we use Table 1 and transform it to a 2-anonymous table using the different match definitions in turn. Table 2 shows the result of the anonymization of Table 1 to a 2-anonymous table. The table has only 3 rows, as all rows of the original table which contain NULL values (rows A, D, and E) had to be removed before the generalization - hence the table is also 3-anonymous. For the rest of this paper we always follow the full-domain generalization scheme [17,24,26] in our examples, however, the considerations are applicable to all algorithms for k-anonymization.

Table 2. 3-anonymity with basic match

	Gender	Height	Job	ZIP	Condition
B	All	All	All	9020	Hepatitis
C	All	All	All	9020	Flu
F	All	All	All	9020	Gastritis

3.4   Extended Match

In extended match NULL values are treated like any other value, in particular, a NULL value only matches with another NULL value but not with any values from the domains of the attributes.

Definition 3 *(Extended match)*. *Let T be a table and Q the set of quasi identifier attributes of T. For two rows $t_1, t_2 \in T$ we define the extended match as*

$$t_1 \sim_e t_2 :\Longleftrightarrow \forall q \in Q : t_1[q] = t_2[q] \vee (t_1[q] isNULL \wedge t_2[q] isNULL)$$

The extended matching definition can be used to extend existing anonymization algorithms. First we have to extend all generalization hierarchies with a branch with the value NULL on each level of the hierarchy below the root 'ALL'. Using these extended hierarchies we can apply the generalization method again and receive Table 3 which is 2-anonymous with respect to the extended match. Note that in contrast to the basic match no row has been lost.

The aim of k-anonymity is to prevent attacks on released data, in particular, record linking attacks [12], i.e. joining a table with some known information

Table 3. 2-anonymity with extended match

	Gender	Height	Job	ZIP	Condition
A	All	All	NULL	9020	Cancer
B	All	All	Admin	9020	Hepatitis
C	All	All	Admin	9020	Flu
D	All	All	Technical	9020	Pneumonia
E	All	All	NULL	9020	Malaria
F	All	All	Technical	9020	Gastritis

to associate values of sensitive attributes with some data owner. In particular, matching any record containing quasi identifiers with the released table should result in no or at least k hits. It is easy to see that this requirement is fulfilled, if any query posed on the released table in the form of "Select * From T where search_condition" yields 0 or at least k result rows, if the search condition only contains predicates on quasi identifiers.

Theorem 1 *(link-safe)*. *Let T be a table and Q a set of quasi identifier attributes and let $\pi_Q T$ be the projection of T on Q. If T is k-anonymous with respect to $\sim_e$ then for all search conditions p the query "Select * From $\pi_Q T$ where p" returns 0 or at least k rows.*

Proof. The theorem follows from the observation that if a row $t \in \pi_Q T$ satisfies the search condition of the query then all rows matching t according to the extended match also satisfy the search condition. Because T is k-anonymous with respect to $\sim_e$ there are at least k such rows.

3.5   Maybe Match

For extending the domain of the match predicate to also consider NULL values we can build on the treatment of NULL values in Codd's *maybe* operations for the relational algebra [6]. The maybe selection operator does not only return those rows for which the selection predicate is satisfied, but also all those rows which satisfy the selection predicate if NULL values are replaced by suitable values.

Applying the concept of 'maybe' selects to the matching of rows, we define the maybe match as follows: NULL matches both with NULL and other values. NULL values in the rows can be used as wildcards in both directions. For an example: The tuples (a, b, c), (a, NULL, c), (NULL, NULL, NULL) all match can be grouped together.

Definition 4 *(maybe match)*. *Let T be a table and Q the set of quasi identifier attributes of T. For two rows $t_1, t_2 \in T$ we define the maybe match as*
$$t_1 \sim_m t_2 :\Longleftrightarrow \forall q \in Q : t_1[q] = t_2[q] \vee (t_1[q] \text{ is NULL} \vee t_2[q] \text{ is NULL}).$$

The maybe match is not transitive and does not lead to an equivalence partitioning of the table (in contrast to basic and extended match). We call a row t and its matching rows the *match-group* of t. k-groups of different rows might overlap without being equal. For an example, the tuple (a, NULL, NULL) matches both with (a, b, c) and with (a, e, f) and so (a, NULL, NULL) is contained in several match-groups.

Applying maybe matching in the generalization method we compute the table shown in Table 4. In this table the match-groups are built from the following match relations $A \sim_m C$, $B \sim_m E$, $D \sim_m E$, $E \sim_m F$, such that each row matches with one other row.

Table 4. 2-anonymity with maybe match

	Gender	Height	Job	ZIP	Condition
A	f	161–180	NULL	9020	Cancer
B	m	181–200	Mayor	9020	Hepatitis
C	f	161–180	Clerk	9020	Flu
D	m	NULL	Technician	9020	Pneumonia
E	m	181–200	NULL	9020	Malaria
F	m	181–200	Pilot	9020	Gastritis

Let us now analyze whether this table is safe. First we try an attack on missing values.

Hampering Reconstruction [22] is an attack that shows how a value that was suppressed in the anonymization process can be reconstructed. We extend it here to cover also missing values in the original table. For example, if an adversary knows that Daniel's data are in the table and Daniel is 205 cm tall, then he can associate row D with Daniel.

Hampering reconstruction requires that a value for some attribute exists in the real world, but is not recorded in the database. It shows that tables, where NULL values have the semantics of missing values, may be compromisable.

Next we show that there are also attacks possible on NULL values, which have the semantics 'not applicable', i.e. for which no value exists in the real world.

We introduce the novel NULL-*identifier attack*, which uses knowledge whether an attribute is applicable for some row. For example, let us assume that an adversary knows that Alice, a female patient, is not employed, and therefore the value in the Job attribute has to be NULL. He can thus query Table 4 with the search condition 'Gender = "f" and Job is NULL' and retrieves row A.

The NULL-identifier attack leverages on the knowledge that a certain value does not exist. Therefore, the row in the table has to have the value NULL in the corresponding attribute. Anonymization based on maybe match is thus not safe against NULL identifier attacks.

With hampering reconstruction and NULL-identifier attack we show that tables which are k-anonymous with respect to maybe match are not safe from linking attacks. Both attacks exploit situations where there are less than *k* NULL values in some attribute within a *k*-group.

In conclusion we observe that generalizations using maybe-match are vulnerable against extended hampering reconstruction (NULLs as 'missing values') and NULL-identifier attack (NULLs as 'attribute not applicable').

3.6 Right Maybe Match

We restrict the definition of the maybe match, such that within a match-group a NULL value in some attribute has to appear at least *k* times, but use the wildcard character of NULL for matching other values with NULL. For an example, (a, b, c) would match with (a, NULL, NULL) and (a, b, NULL) but not vice versa. The motivation for this is to enforce the existence of *k* NULL values in some attribute within a match-group to avoid hampering reconstruction and NULL-identifier attacks. In our example, (a, b, c) matches with the two other tuples, (a, b, NULL) with one other tuple. The expectation was that a single tuple like (a, b, c) demanding further generalization with extended match would match with tuples containing NULL values and therefore would not require further generalization.

Definition 5 *(right maybe match). Let T be a table and Q the set of quasi identifier attributes of T. For two rows $t_1, t_2 \in T$ we define the right maybe match as*

$$t_1 \sim_r t_2 :\Longleftrightarrow \forall q \in Q : t_1[q] = t_2[q] \vee t_2[q] is\text{NULL}.$$

The right maybe match relation is reflexive and transitive, but not symmetrical and does not define an equivalence partitioning on a table. The non-symmetry is viable, as the definition of *k*-anonymity requires that each tuple matches with *k* other tuples, however, it does not require that the match relation defines equivalence classes.

The result of anonymization with right maybe match is shown in the example of Table 5. Here the following match relations are found: A $\sim_r$ E, B $\sim_r$ A, B $\sim_r$ E, C $\sim_r$ A, C $\sim_r$ E, D $\sim_r$ A, D $\sim_r$ E, E $\sim_r$ A, F $\sim_r$ A, F $\sim_r$ E.

Table 5. 2-anonymity with right maybe match

	Gender	Height	Job	ZIP	Condition
A	All	All	NULL	9020	Cancer
B	All	All	Mayor	9020	Hepatitis
C	All	All	Clerk	9020	Flu
D	All	All	Technician	9020	Pneumonia
E	All	All	NULL	9020	Malaria
F	All	All	Pilot	9020	Gastritis

Now let us analyze, whether the right maybe match admits tables which are safe. We first observe that there are queries on the quasi identifier projection of a table which yield less than k results. For example, the search condition 'Job = "technician"' would only return row D. However, this query might lead to a wrong result for an adversary, because it is possible, that the 'true' value in the Job attribute is also "technician" for the rows A and E but just missing. Therefore, a rational adversary would use a maybe query instead ('Job = "technician" or Job is NULL').

It is possible to show that right maybe match is safe for all maybe queries.

We introduce the *singularity attack* to show that straight (not maybe) queries make sense and possibly compromise the data. A singularity attack uses knowledge that some value is unique, at least for some combinations of other attribute values. For example, ZIP code and Job may be compromised, when the job = "Mayor" and there is never more than one mayor in a town (i.e. per ZIP code). In such cases an adversary will use straight queries rather than maybe queries and can compromise tables which are k-anonymous with respect to the right maybe match.

The singularity attack shows that tables which are k-anonymous with respect to right maybe match are not safe against attacks. The singularity attack does not depend on the type of NULL values (non-applicable, missing, no-information), such that we have to dismiss anonymization based on the right-maybe match.

3.7   An Extended Generalization Algorithm

A detailed analysis of NULL values and their matching operators allows to extend generalization algorithms (see Sect. 2) with minimal effort to cover also NULL values in the input table using the k-anonymity definition with extended match. There are only two extensions necessary: (1) the generalization hierarchies of each quasi identifier is extended with an additional branch below the root that contains a NULL value in each level of the hierarchy. (2) k-anonymity is tested with extended match. With these extensions, the anonymization algorithms can accept microdata with NULL values without any preprocessing. It is easy to see that the complexity of the algorithms is not changed by this extension.

Figure 2 shows, how the generalization hierarchies of our running example were extended, in order to apply extended matching. Note that for consistency reasons, there has to be a NULL value on each level of the hierarchy.

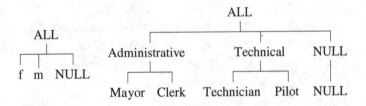

Fig. 2. Generalization hierarchies for NULL value handling in the running example

The extended generalization algorithm as well as the extended match are implemented in the tool ANON that is introduced in the next section. ANON was used to execute experiments on microdata with NULL values to analyze whether the explicit treatment of NULL values actually reduces information loss. The results are shown in Sect. 5.

4   The Anonymization System ANON

We implemented a flexible and customizable tool, called ANON [3], for computing k-anonymous and ℓ-diverse tables based on anonymization by generalization and tuple suppression [26,30] where the information loss can be defined by the users explicitly through penalties in the generalization hierarchies and through priorities for attribute generalizations [27,28]. The contribution of ANON is on one hand the computation of a k-anonymous and ℓ-diverse generalization of a given table with minimal information loss, where this information loss can be defined application specific in a fine grained way when defining the generalization hierarchies [3] and on the other hand the explicit treatment of NULL values. We implemented both anonymization with basic match and with extended match such that ANON offers two ways of handling NULL values: removing rows with NULL values before anonymization (basic match), or treating NULL values as any other value (extended match). We did not implement anonymization with maybe match or right maybe match due to their vulnerability as shown in Sect. 3.

4.1   User-Specific Requirements and Information Loss

ANON aims at adapting to the application specific data requirements by allowing the user to customize the anonymization procedure by defining application specific information loss calculation. The motivation for this is the observation that the requirements for the precision of attribute values vary enormously between different applications. For an example, in one application the age is needed in fine granularity while in a different application the body mass index is needed in detail and age is sufficient in 10-year intervals. So we argue that technical information loss definitions like those based on Shannon's definition of entropy cannot reflect the usability of a data set for a specific application. For steering the search for an optimal solution the users may specify priorities for the generalization of quasi identifiers, to specify an information loss for each generalization step and to set generalization limits. The calculation of information loss is used in a best-first search to determine the best anonymized table with generalization and tuple suppression.

Weighted information loss is calculated with the formula

$$\text{WIL} = \sum_{i=1}^{n} prios[i] * \varphi_{\alpha_i}^{levels[i]},$$

where n is the number of quasi identifiers, $\varphi_{\alpha_i}^{levels[i]}$ the loss of information if the attribute α_i is generalized to the level $levels[i]$, and $prios[i]$ is the priority of the attribute α_i.

With this formula, the user can influence the information loss with the following user-specified information:

Attribute priorities. A User can assign a particular priority to each quasi identifier attribute. These priorities influence the computation of generalization loss. The intention is that attributes with a high priority are generalized to lower levels than attributes with lower priorities.

Generalization limits. If the values are useful only up to a particular generalization level, the user can limit the generalization of an attribute to this level.

Generalization penalties. Generalization penalties define the information loss for each generalization level. The top level of a generalization hierarchy has an information loss of 100 %.

This user-specified information and its impact in the search algorithm guarantee that the user will obtain such results that suit the user's requirements in the best possible way.

4.2 Architecture and Implementation

ANON is implemented in Java and is available in two distributions: as a Web Service and as an executable platform-independent java archive (JAR) with a simple graphical user interface. The anonymization is controlled by the ANON definition file, which is a combination of an anonymization settings file and a metadata file. As shown in Fig. 3, the ANON definition file is the main ANON input that determines the microdata source(s) and the outputs, as well as the anonymization process itself.

ANON definition is an XML file that consists of the following five fundamental sections:

Parameters define the anonymization settings (the value of k, maximal suppression threshold max_supp, type of the search algorithm, ANON report settings and missing value handling details).

Datasource definition defines the source(s) of data that should be anonymized (database(s), XML or CSV-file(s)).

Output definition defines the target where the anonymized data should be saved (database, XML or CSV-file).

Attributes definition defines which attributes should be read from the source data and how these should be handled. Each quasi identifier attribute (k-attribute) should have assigned its priority and maximal generalization limit. Each sensitive attribute (l-attribute) must be configured with the ℓ-diversity type that should be used for checking, and its desired parameter values (ℓ).

Generalization hierarchies define value generalization hierarchies that are used for anonymization. For every quasi identifier attribute (k-attribute) there should be one generalization hierarchy, upon which the values are

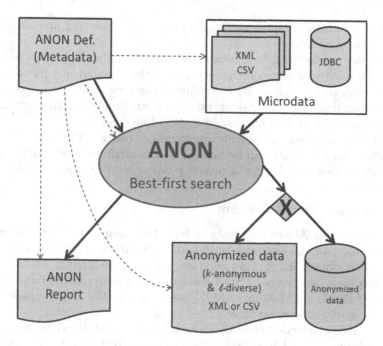

Fig. 3. ANON architecture

generalized. Each generalization hierarchy contains information about the hierarchy levels inclusive their information loss and a value generalization tree. This tree must contain all the values that the corresponding quasi identifier can hold in the microdata.

ANON is capable of anonymizing data from multiple sources that have one of the following formats: database connection (JDBC), XML-file or CSV-file. The result can be saved as one of these formats as well. Besides the anonymization outcome, user can decide to receive an ANON report, which informs about the anonymization process and eventual failures.

ANON offers the user the possibility to select the attributes that the user wants to handle in the anonymization process and mark them with one of the following anonymization types:

- k-attribute,
- l-attribute,
- dontcare,
- ignore.

The attributes that should be skipped from the result should be marked with "ignore". Alternatively they can be left out of the ANON definition file to raise the same effect. If an attribute does not play any role for the individuals privacy and should appear unchanged in the result, then it should be marked as "dontcare". The remaining two types are those that are relevant for the anonymization process.

Attributes marked with "k-attribute" are quasi identifiers that must be transformed to a particular generalization level, such that k-anonymity for the whole table is satisfied. For this kind of attributes, the user should also specify the generalization limit and attribute priority. For l-attributes (sensitive attributes), the ℓ-diversity type and its parameter(s) should be defined.

ANON is designed to provide anonymized tables with multi-attribute ℓ-diversity. The ℓ-diversity can be defined on the attribute level. Furthermore, ANON allows to assign different ℓ-diversity types to different attributes.

4.3 Anonymization Algorithm

The anonymization Algorithm 1 features both basic match and extended match for computing partitions defined by the quasi identifiers. The variables used in ANON's Partitioning Algorithm 1 and their meaning are listed in Table 6. After a table has been partitioned, ANON checks whether each partition has at least k tuples. If this is not the case, ANON uses generalization and suppression to generate a more coarse grain table.

ANON's anonymization algorithm uses best-first search algorithm to find the optimal solution and weighted information loss described in Sect. 4.1 to evaluate the cost of a potential solution (generalized table).

The algorithm consists of 2 main parts: table search (function ANONYMIZE-TABLE - see Algorithm 2) and privacy test (function PRIVACY-TEST - see Algorithm 3). Required input parameters of both functions of Algorithms 2 and 3 are listed and described in Table 7. Furthermore, other variables and values are listed and described in Table 8, instance variables of a node in Table 9 and functions in Table 10.

ANON is customizable, so the implementation offers an abstract class of search algorithms that can easily be extended by new search algorithms. Similar interfaces are provided for information loss calculation, as well as for ℓ-diversity check. Details about the implementation are given in the next section.

Table 6. Variables used in ANON's Partitioning Algorithm 1

Variable	Description
null_handling	Denotes if NULL values are handled or not. If *null_handling* is TRUE then the partitioning algorithm will use the extended match, otherwise the basic match
partition	A partition is a set of tuples that match each other
partitionset	Set of partitions
table	A table with microdata
tuple	A table row
matched	Denotes if a tuple matches a partition

Algorithm 1. ANON's Partitioning Algorithm

Input: *table, null_handling*
Output: set of partitions *partitionset*, where each partition contains tuples with identical values of quasi identifiers or NULLs instead of a quasi identifier value. Partitions are disjoint.

```
 1: function PARTITION-TABLE(table, null_handling)
 2:     partitionset ← { }
 3:     for each tuple in table do
 4:         matched ← FALSE
 5:         for each partition in partitionset do
 6:             if tuple matches partition then
 7:                 add tuple into partition
 8:                 matched ← TRUE
 9:             end if
10:         end for
11:         if matched = FALSE then
12:             if tuple does not contain NULL or null_handling = TRUE then
13:                 partition ← {tuple}
14:                 add partition into partitionset
15:             end if
16:         end if
17:     end for
18:     return partitionset
19: end function
```

Table 7. Input parameters required by the function ANONYMIZE-TABLE

Parameter	Description
table	Original table that has to be anonymized
limits	Array with generalization level limits for all quasi identifiers
prios	Array with priorities for all quasi identifiers
k_param	*k* - the minimal partition size
l_params	Array with minimal required diversities for all sensitive attributes
max_supp	Number of tuples that are allowed to be suppressed
null_handling	Denotes if NULL values are handled or not. If *null_handling* is TRUE then the partitioning algorithm will use the extended match, otherwise the basic match

4.4 Performance Analysis

ANON's anonymization algorithm (a best-first search instantiation) is optimal and complete. If generalization limits are set to less than the number of generalization levels, it is possible that the algorithm will not find a solution (because there is none). If no limits are set, it always finds a solution, which is in worst case a completely generalized table.

Table 8. Variables and values used in ANON's Priority-Based Algorithms 2 and 3

Variable/value	Description
open	List of potential solutions (generalized tables)
visited	List of potential solutions that have already been added to *open*
best	Generalized table from *open* with the lowest information loss
levels	Generalization levels of a potential solution
child	Child node - a potential solution with a table generalized to the next higher level at one quasi identifier, while the other attributes' levels remain the same as the levels of the parent
supp_tuples	Number of tuples from all partitions that violate k-anonymity and/or ℓ-diversity
partition	Table partition - a set of records with the same quasi identifier values
diversities	Array with diversities of one partition for all sensitive attributes
NIL	Represents a NULL value
FAILURE	Denotes that an anonymized table, which satisfies all constraints, could not be found

Table 9. Instance variables of a node used in ANON's Priority-Based Algorithms 2 and 3

Variable	Description
node.PARENT	Parent node - potential solution from which *node* was deduced
node.LEVELS	Generalization levels of *node*.TABLE - represents action
node.WIL	Weighted information loss - represents total path cost
node.TABLE	Table with values generalized to *node*.LEVELS - represents state

As all optimizing generalization algorithms ANON has in the worst case exponential time complexity caused by the state space, which grows exponentially with the number of quasi identifier attributes and their limits. However, for a given set of quasi identifiers it scales nicely for increasing sizes of the data set. The set of experiments presented in this section were intended to analyze whether the anonymization with customizable calculation of information loss is feasible for large real-world datasets.

We did not compare ANON with other algorithms. The intention of the experiments was not to show that ANON reduces information loss in general, as other algorithms do not feature an application specific calculation of information loss so any such comparison would be pointless. We also do not claim that ANON is the fastest anonymization algorithm and the system could be accelerated e.g. by applying other heuristic search procedures.

The microdata used for the experiments was the *Adult Data Set* from UCI Machine Learning Repository [11] commonly used for performance experiments

Table 10. Functions used in ANON's Priority-Based Algorithms 2 and 3

Function	Description
ANONYMIZE-TABLE	Input: *table, limits, prios, k_param, max_supp* Output: Anonymized table, which satisfies all input constraints, or FAILURE if no solution could be found
PARTITION-TABLE	Input: *table, null_handling* Output: set of partitions *partitionset*, where each partition contains tuples with identical values of quasi identifiers or NULL instead of a quasi identifier value.
PRIVACY-TEST	Input: *node, k_param, max_supp* Output: TRUE if *node*.TABLE satisfies *k*-anonymity and ℓ-diversity, else FALSE
MAKE-NODE	Input: *table* Output: Node that represents *table* in the search space with LEVELS initialized with 1.
GENERALIZE	Input: *table, node*.LEVELS Output: Table with values generalized to node.LEVELS.
CHILD-NODE	Input: *parent, levels, prios* Output: Node with PARENT = *parent*, LEVELS = *levels*.
CALCULATE-WIL	Input: *levels, prios* Output: Weighted information loss of a generalized table, calculated as follows: $$\sum_{i=0}^{length(levels)-1} prios[i] * \varphi_{\alpha_i}^{levels[i]} \quad \varphi_{\alpha_i}^{levels[i]}$$ is the loss of information if the attribute α_i gets generalized to the level *levels*[*i*].
COUNT-TUPLES	Input: *partition* Output: Number of tuples in the *partition*.
length	Input: *array* Output: Length of the *array*.

in the microdata privacy literature. The Adult Data Set contains real data collected by the U.S. census bureau in the year 1994. This data is split in a training set and a test set. For our experiments we merged both sets together and tuples with unknown values were removed. After data cleaning, the set contained 45,222 tuples. To provide comparable results, the same data preparation was undertaken as described in [17,18]. From the 15 attributes in the data set, the identical nine were chosen as in [18]. As shown in Table 11, the attributes *age, gender, race, marital status, education, native country* and *workclass* were used as quasi identifiers and the attributes *salary class* and *occupation* were used as sensitive attributes. Generalization hierarchies for the used quasi identifiers were constructed in a semantically logical way.

There were two experiment runs, each with almost 800 anonymizations: one with the use of priorities and information loss and one without them, to imitate the optimal search algorithms without a cost function (e.g. MinGen).

Algorithm 2. ANON's Priority-based Algorithm - Part 1 (Search Algorithm)

Input: *table, limits, prios, k_param, l_params, max_supp, null_handling*
Output: anonymized table satisfying k-anonymity and ℓ-diversity or FAILURE if no solution could be found

```
 1: function ANONYMIZE-TABLE(table, limits, prios, k_param, l_params, max_supp)
 2:     open ← {MAKE-NODE(table)}
 3:     visited ← { }
 4:     while open is not empty do
 5:         best ← node n in open with the lowest n.WIL value
 6:         if best.TABLE = NIL then
 7:             best.TABLE ← GENERALIZE(table, best.LEVELS)
 8:         end if
 9:         if     PRIVACY-TEST(best, k_param, l_params, max_supp, null_handling)
    then
10:             return best.TABLE
11:         else                                              ▷ expand best
12:             for i ← 0 to length(limits) − 1 do
13:                 levels ← best.LEVELS
14:                 if limits[i] > levels[i] then
15:                     levels[i] ← levels[i] + 1
16:                     child ← CHILD-NODE(best, levels, prios)
17:                     if child not in visited then
18:                         add child into open
19:                         add child into visited
20:                     end if
21:                 end if
22:             end for
23:             best.TABLE ← NIL
24:             remove best from open
25:         end if
26:     end while
27:     return FAILURE
28: end function

29: function CHILD-NODE(parent, levels, prios)
30:     return a node with
31:             PARENT ← parent,
32:             LEVELS ← levels,
33:             WIL ← CALCULATE-WIL(levels, prios)  ▷ Weighted Information Loss
34:             TABLE ← NIL
35: end function
```

Information loss values are listed within generalization hierarchies that come with ANON. The priority order of attributes (starting with a low priority) in the first run was {*age, native country, education, marital status, workclass, race, sex*}. In the second run (without the cost function), an implicit priority order was derived from the attributes order in the ANON definition file, which was

---

Algorithm 3. ANON's Priority-based Algorithm - Part 2 (Privacy Test)

```
36: function PRIVACY-TEST(node, k_param, l_params, max_supp, null_handling)
37:     supp_tuples ← 0
38:     for each partition in PARTITION-TABLE(node.TABLE, null_handling) do
39:         if COUNT-TUPLES(partition) ≥ k_param then        ▷ k-anonymity satisfied
40:             diversities ← CALCULATE-DIVERSITIES(partition)
41:             for i ← 0 to length(l_params) − 1 do
42:                 if diversities[i] < l_params[i] then       ▷ ℓ-diversity not satisfied
43:                     remove partition from node.TABLE
44:                     supp_tuples ← supp_tuples + COUNT-TUPLES(partition)
45:                     break
46:                 end if
47:             end for
48:         else                                              ▷ k-anonymity not satisfied
49:             remove partition from node.TABLE
50:             supp_tuples ← supp_tuples + COUNT-TUPLES(partition)
51:         end if
52:         if supp_tuples > max_supp then                    ▷ privacy not satisfied
53:             return FALSE
54:         end if
55:     end for
56:     return TRUE                        ▷ privacy satisfied (supp_tuples ≤ max_supp)
57: end function
```

{*age, sex, race, marital status, education, native country, workclass*}. Generalization limits were not set (they equaled to the no. of generalization levels) to avoid an anonymization without a solution.

The experiments were performed to estimate the "real case" complexity and the impact of different parameters on the number of visited nodes, resulting information loss and average partition size. These parameters are listed in Table 12.

Table 11. Adult Data Set description (adapted from [18])

	Attribute	Domain size	Generalization type	No. of gen. levels
1	Age	74	Ranges (5, 10, 20, 100)	4
2	Gender	2	Taxonomy tree	1
3	Race	5	Taxonomy tree	1
4	Marital Status	7	Taxonomy tree	2
5	Education	16	Taxonomy tree	3
6	Native Country	41	Taxonomy tree	2
7	Work Class	7	Taxonomy tree	2
8	Salary class	2	Sensitive att	
9	Occupation	14	Sensitive att	

Table 12. Experiments' parameters and their values

Parameter	Chart notation	Values
n	QID	1, 2, 3, 4, 5, 6, 7
k	k	2, 3, 5, 10, 14, 20, 100, 200, 1000
ℓ_{α_8}	l1	1, 2
ℓ_{α_9}	l2	1, 2, 3, 4, 5, 6, 7, 8, 9, 10, 11, 12, 13, 14
max_supp	supp	0 %, 1 %, 10 %
cost function (WIL)	With priorities	Used (first run),
	W/o priorities	not used (second run)

There were over 1,500 anonymizations produced, where the parameters were set to almost all possible value combinations.

The first two experiments deal with the time complexity. In these experiments, the impact of the quasi identifiers' increase on the number of visited nodes and anonymization time was analyzed. Both experiments were executed with four different anonymization settings groups: (1) $k = 2$ with 0 % tuple suppression (black line with white markers), (2) $k = 2$ with 1 % suppression limit (black line with black markers), (3) $k = 10$ with 0 % suppression and $\ell_{\alpha_8} = 2$, $\ell_{\alpha_9} = 10$ (gray line with white markers) and (4) $k = 10$ with 1 % suppression and $\ell_{\alpha_8} = 2$, $\ell_{\alpha_9} = 10$ (gray line with black markers). The dotted line denotes the maximal possible number of visited nodes ($\Pi_{i=1}^{n} limits[i]$) and the approximated maximal required anonymization time, respectively. As approximation, the anonymization settings with $k = 14$, 0 % suppression and $\ell_{\alpha_8} = 2$, $\ell_{\alpha_9} = 14$ were used. These settings were noticed to result in maximal possible values, because only the last queued node with completely generalized table satisfies these settings.

Time complexity of k-anonymity algorithms similar to ANON grows with the number of quasi identifiers [5,12,17]. In contrast to that, the number of tuples in a table does not have a big impact on time complexity. Table size is just a constant factor multiplied by the number of nodes, which does not affect the number of visited nodes itself and can therefore be neglected. Figure 4 confirms for ANON that time complexity grows with the increase in quasi identifiers.

Figure 5 represents the same experiment, where, instead of number of visited nodes, the time was measured. If we compare both figures, it is easy to see that time depends on the number of nodes and some other factors like generalization hierarchy height. If we observe the ℓ-diversity lines (gray lines) in Figs. 4 and 5, we can notice that these lines have a slightly higher slope in the chart with time on the y-axis than in the chart with nodes on the y-axis. The explanation for this phenomenon is hidden in diversity calculation effort. ANON does not need to calculate the size of a partition (relevant for k-anonymity checking) explicitly, because it is managed together with a partition. The diversity (relevant for ℓ-diversity checking), in opposition to partition size, has to be calculated extra for each partition, if the partition is k-anonymous.

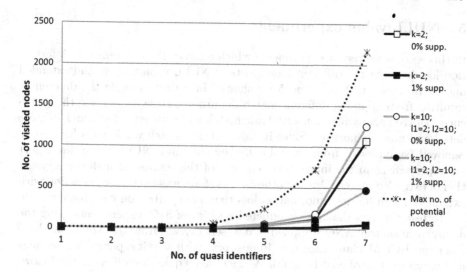

Fig. 4. Search algorithm complexity (number of nodes)

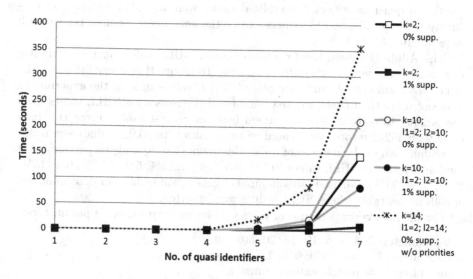

Fig. 5. Search algorithm complexity (time)

The charts Figs. 4 and 5 show the highly significant impact of tuple suppression. Both black lines have the same settings, except the maximal suppression limit (white markers 0 %, black markers 1 %). However, the difference in the results is huge. It took 1,075 nodes to anonymize 7 quasi identifiers to a 2-anonymous table with no suppression (black line with white markers) and with just 1 % suppression (max. 453 tuples may be eliminated), it took only 60 nodes to find the optimal solution. More experiment results and analysis of different parameters' impact can be found in [3].

5   NULL value experiments

In this section we report experiments which analyze the differences in information loss between the different treatments of NULL values: basic and extended match. In Tables 1, 2, 3, 4 and 5 we showed in a tiny example the differences resulting from applying different match definitions. Here we analyze the differences between anonymization with basic match and with extended match in more detail and more exhaustive. Since in case of basic match whole rows have to be removed (suppressed) from the table if they contain a NULL value, we expect an improvement in the information content of the results through the application of extended match. Our hypothesis, therefore, was that anonymization with extended match has less information loss than anonymization with basic match.

To test the hypothesis we conducted a series of 3168 experiments using the anonymization tool ANON applied to datasets derived from the Adult Database from the UCI Machine Learning Repository, with varying parameter settings and varying ratios of NULL values. We included 8 quasi identifiers in the following order: *age, sex, race, marital_status, workclass, education, native_country, occupation*. According to this order, tables with $n < 8$ quasi identifiers contain the first n quasi identifiers. Generalized values were provided with help of taxonomy trees for all quasi identifiers except the *age*, where we used ranges with steps $\{5, 10, 20, 100\}$.

The Adult Database itself contains several NULL values in the attributes *workclass, occupation* and *native_country*. To assure that every table has the right target amount of randomly placed NULL values used in the experiments (and not more than that), we first eliminated the rows with NULL values from the original table to obtain a common base for all test tables. From this base table with 45222 records we created 88 test tables with NULL values as a result of combination of the number of quasi identifiers (1 to 8) and the percentage of randomly inserted NULL values (0.1 %, 0.5 %, 1 %, 2.5 %, 5 %, 7.5 %, 10 %, 15 %, 20 %, 25 %, 30 %). We used random number generation in Java for determination of cells in the table where NULL values were inserted.

For the anonymization runs of the 88 tables we used following parameters:

- k-parameter: 2, 3, 4, 5, 10, 15, 20, 50, 100
- max. allowed suppression: 0 %, 1 %
- matching: basic match, extended match.

The max. allowed suppression specifies the quota for suppressing rows (to avoid adverse effects of outliers). For basic match this is in addition to the rows with NULL values, which are removed in a preprocessing step.

Information loss in the anonymization algorithm is caused by row suppression (fraction of rows being suppressed) and by generalization (generalization level of the attributes). For the following experiments the information loss was calculated with the following formula:

$$IL = \frac{s}{n} + \sum_{i=1}^{m} \frac{gl_{\alpha_i}}{ht_{\alpha_i}} \times \frac{1}{m} \times \frac{n-s}{n}$$

n Number of rows
s Number of removed rows
m Number of quasi identifier attributes
gl_{α_i} Generalization level of the attribute α_i in the generalized table
ht_{α_i} Height of the generalization hierarchy of the attribute α_i
    (Number of total generalization levels of α_i).

Information loss due to row suppression is represented in the first summand as the fraction of removed tuples. The information loss caused by generalization is calculated as the weighted sum of information losses of all attributes multiplied with the fraction of non-removed tuples. The information loss of an attribute is defined as fraction of the generalization level of this attribute by the number of levels, where the most general level (which carries no information at all)is h and the level in the original table is 0. So if an attribute is generalized to the level 3 of a 5 level hierarchy we define the information loss a 3/5. Note that the values of an attribute in all tuples are generalized to the same level as we apply the global recording strategy. Therefore we can calculate the information loss at the attribute level.

We show first representative comparisons of the information loss between basic match and extended match without and with row suppression. In the figures results of extended match is shown in light gray bar and those of basic match in black bars. Each bar represents the information loss of one anonymization run.

Figures 6 and 7 show anonymizations without row suppression (max. supp. = 0 %). Anonymizations with extended match (light gray bars) tend to have constant information loss, whereas the information loss of anonymizations with basic match (black bars) has a growing trend with increasing percentage of NULL values. This behavior can be explained with 2 influence factors: (1) number of

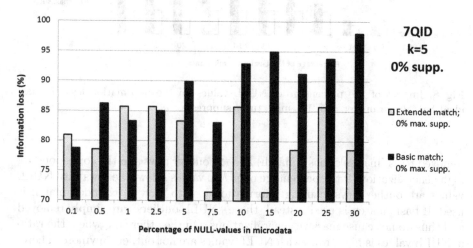

Fig. 6. Impact of the percentage of NULL values on the information loss (7 quasi identifiers, *k*-parameter = 5, 0 % max. row suppression).

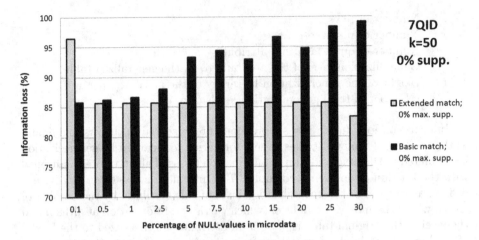

Fig. 7. Impact of the percentage of NULL values on the information loss (7 quasi identifiers, k-parameter $= 50$, 0% max. row suppression).

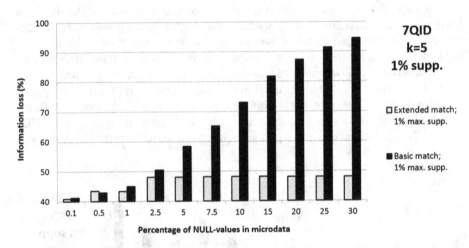

Fig. 8. Impact of the percentage of NULL values on the information loss (7 quasi identifiers, k-parameter $= 5$, 1% max. tuple suppression).

removed rows (in case of basic match) and (2) outlier rows and the corresponding high generalization. If the percentage of NULL values is low, the rows with NULL values are outlier rows, causing information loss to grow if extended match is used. If basic match is used instead, those "NULL-outliers" are simply removed and thus do not cause massive generalizations. On the other end, where the ratio of NULL values is high, rows with NULL values are not outliers anymore. Therefore, they do not increase the information loss if extended match is used. For basic match, however, information loss is proportional to the ratio of rows with

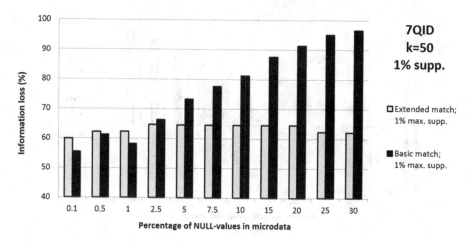

Fig. 9. Impact of the percentage of NULL values on the information loss (7 quasi identifiers, *k*-parameter = 50, 1 % max. tuple suppression).

NULL values leading to an increase in information loss with increasing ratios of NULL values.

Figures 8 and 9 show anonymizations with the same setup as those in Figs. 6 and 7, but with row suppression of up to 1 %. Here extended match is no longer so sensible on NULL outliers and results in an almost constant information loss over increasing ratio of NULL values (light gray bars), while information loss grows drastically for basic match (black bars). That for low ratios of NULL values (below 1 %) basic match is slightly better than extended match might be due that for basic match more rows are removed (number of rows with NULL plus 1 % of the rows without NULL).

Figure 10 shows the aggregated results of all 3168 anonymizations in our experiment. Each bar represents the average difference in information loss of anonymizations with basic match and anonymizations with extended match, calculated over all 8 quasi identifiers and all 9 different *k*-parameters. The light gray bars represent the setups without row suppression (max. supp. = 0 %) and the dark gray bars the setups with 1 % max. suppression.

To summarize the results: The experiments showed that the best method in general was extended match with 1 % row suppression. For very low ratios of NULL values basic match was slightly better. Extended match without row suppression performs worse for low ratios of NULL values, because it suffers from the generalizations caused by NULL outliers. Basic match was only favorable for very low numbers of NULL values and the quality of the results deteriorates with increasing ratios of NULL values, caused by the removal of all rows with NULL values. Furthermore, the information loss for extended match with row suppression did not seem to be influenced by the number of NULL values in the data set, as shown by the almost constant information loss over varying ratios of NULL values.

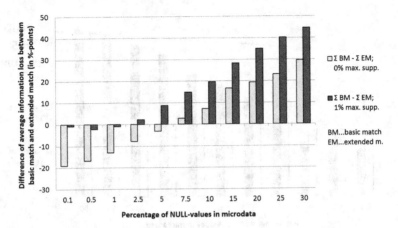

Fig. 10. Average advantage in %-points of information loss of the extended match over the basic match, depending on the percentage of NULL values in a table. In the positive y-area the extended match outperforms the basic match.

6   Conclusions

NULL values (missing values, not applicable attributes) appear frequently in microdata. Surprisingly, current anonymization algorithms require that all rows containing NULL values are removed from a table before it can be anonymized. We analyzed the effects of including NULL values in the definition of k-anonymity in detail and showed that the extended match where NULL values match (only) with other NULL values is a correct approach for extending k-anonymity to cover missing values. We introduced two new attacks that show that a further relaxation of the match operator which interprets NULL values as wildcards in the sense of Codd's maybe select leads to tables which can be attacked successfully. The extension of k-anonymity to tables with NULL values reduces the information loss induced by the removal of rows with NULL values by current anonymization algorithms and avoids the introduction of biases. Experiments showed that extended match reduces information loss for a generalization algorithm with row suppression considerably. The definition of k-anonymity we propose here can be used easily as basis for extending other anonymization algorithms to also cover tables with NULL values in an adequate and save way.

References

1. Aggarwal, G., Feder, T., Kenthapadi, K., Motwani, R., Panigrahy, R., Thomas, D., Zhu, A.: Approximation algorithms for k-anonymity. In: Proceedings of the International Conference on Database Theory, ICDT 2005 (2005)
2. Bayardo, R.J., Agrawal, R.: Data privacy through optimal k-anonymization. In: Proceedings of the 21st International Conference on Data Engineering, ICDE 2005, pp. 217–228 (2005)

3. Ciglic, M., Eder, J., Koncilia, C.: ANON - a flexible tool for achieving optimal *k*-anonymous and ℓ-diverse tables. Technical report, University of Klagenfurt (2014). http://isys.uni-klu.ac.at/PDF/2014-ANON-Techreport.pdf
4. Ciglic, M., Eder, J., Koncilia, C.: *k*-anonymity of microdata with NULL values. In: Decker, H., Lhotská, L., Link, S., Spies, M., Wagner, R.R. (eds.) DEXA 2014, Part I. LNCS, vol. 8644, pp. 328–342. Springer, Heidelberg (2014)
5. Ciriani, V., De Capitani di Vimercati, S., Foresti, S., Samarati, P.: *k*-anonymity. In: Yu, T., Jajodia, S. (eds.) SDMDS 2007. AISC, vol. 33, pp. 323–353. Springer, New York (2007)
6. Codd, E.F.: Extending the database relational model to capture more meaning. ACM Trans. Database Syst. **4**(4), 397–434 (1979)
7. Cox, L.H.: Suppression methodology and statistical disclosure control. J. Am. Stat. Assoc. **75**(370), 377–385 (1980)
8. Eder, J., Dabringer, C., Schicho, M., Stark, K.: Information systems for federated biobanks. In: Hameurlain, A., Küng, J., Wagner, R. (eds.) Transactions on Large-Scale Data- and Knowledge-Centered Systems I. LNCS, vol. 5740, pp. 156–190. Springer, Heidelberg (2009)
9. Eder, J., Gottweis, H., Zatloukal, K.: IT solutions for privacy protection in biobanking. Public Health Genomics **15**(5), 254–262 (2012)
10. Eder, J., Stark, K., Asslaber, M., Abuja, P.M., Gottweis, H., Trauner, M., Mischinger, H.J., Schippinger, W., Berghold, A., Denk, H., Zatloukal, K.: The genome austria tissue bank. Pathobiology **74**(4), 251–8 (2007)
11. Frank, A., Asuncion, A.: UCI machine learning repository (2010). http://archive. ics.uci.edu/ml
12. Fung, B.C.M., Wang, K., Fu, A.W.-C., Yu, P.S.: Introduction to Privacy-Preserving Data Publishing: Concepts and Techniques, 1st edn. Chapman & Hall/CRC, Boca Raton (2010)
13. Gaskell, G., Gottweis, H., Starkbaum, J., Gerber, M.M., Broerse, J., Gottweis, U., Hobbs, A., Ilpo, H., Paschou, M., Snell, K., Soulier, A.: Publics and biobanks: Pan-European diversity and the challenge of responsible innovation. Eur. J. Hum. Genet. **21**(1), 14–20 (2013)
14. ISO: ISO/IEC 9075-2:2011 Information technology – Database languages – SQL – Part 2: Foundation (SQL/Foundation), December 2011
15. Iyengar, V.S.: Transforming data to satisfy privacy constraints. In: Proceedings of the Eighth ACM SIGKDD International Conference on Knowledge Discovery and Data Mining, KDD 2002, pp. 279–288 (2002)
16. Kifer, D., Gehrk, J.: Injecting utility into anonymized datasets. In: Proceedings of the 2006 ACM SIGMOD International Conference on Management of data, SIGMOD 2006, pp. 217–228 (2006)
17. LeFevre, K., DeWitt, D.J., Ramakrishnan, R.: Incognito: efficient full-domain *k*-anonymity. In: Proceedings of the 2005 ACM SIGMOD International Conference on Management of data, SIGMOD 2005, pp. 49–60 (2005)
18. Machanavajjhala, A., Kifer, D., Gehrke, J., Venkitasubramaniam, M.: L-diversity: privacy beyond *k*-anonymity. ACM Trans. Knowl. Disc. Data (TKDD) **1**(1), 3 (2007)
19. Matthews, G.J., Harel, O.: Data confidentiality: a review of methods for statistical disclosure limitation and methods for assessing privacy. Stat. Surv. **5**, 1–29 (2011)
20. Meyden, R.: Logical approaches to incomplete information: a survey. In: Chomicki, J., Saake, G. (eds.) Logics for Databases and Information Systems. The Springer International Series in Engineering and Computer Science, vol. 436, pp. 307–357. Springer, New York (1998). Chapter 10

21. Meyerson, A., Williams, R:. On the complexity of optimal k-anonymity. In: Proceedings of the Twenty-Third ACM SIGMOD-SIGACT-SIGART Symposium on Principles of Database Systems, PODS 2004, pp. 223–228 (2004)

22. Ohrn, A., Ohno-Machado, L.: Using boolean reasoning to anonymize databases. Artif. Intell. Med. **15**(3), 235–254 (1999)

23. Park, H., Shim, K.: Approximate algorithms for k-anonymity. In: Proceedings of the 2007 ACM SIGMOD International Conference on Management of data, SIGMOD 2007, pp. 67–78 (2007)

24. Samarati, P.: Protecting respondents' identities in microdata release. IEEE Trans. Knowl. Data Eng. (TKDE) **13**(6), 1010–1027 (2001)

25. Samarati, P., Sweeney, L.: Generalizing data to provide anonymity when disclosing information (abstract). In: Proceedings of the Seventeenth ACM SIGACT-SIGMOD-SIGART Symposium on Principles of Database Systems, PODS 1998, p. 188 (1998)

26. Samarati, P., Sweeney, L.: Protecting privacy when disclosing information: k-anonymity and its enforcement through generalization and suppression. Technical report (1998)

27. Stark, K., Eder, J., Zatloukal, K.: Priority-based k-anonymity accomplished by weighted generalisation structures. In: Tjoa, A.M., Trujillo, J. (eds.) DaWaK 2006. LNCS, vol. 4081, pp. 394–404. Springer, Heidelberg (2006)

28. Stark, K., Eder, J., Zatloukal, K.: Achieving k-anonymity in datamarts used for gene expressions exploitation. J. Integr. Bioinform. **4**(1), 57 (2007)

29. Sun, X., Wang, H., Li, J., Truta, T.M.: Enhanced p-sensitive k-anonymity models for privacy preserving data publishing. Trans. Data Priv. **1**(2), 53–66 (2008)

30. Sweeney, L.: Achieving k-anonymity privacy protection using generalization and suppression. Int. J. Uncertainty Fuzziness Knowl.-Based Syst. **10**(5), 571–588 (2002)

31. Terrovitis, M., Mamoulis, N., Kalnis, P.: Local and global recoding methods for anonymizing set-valued data. VLDB J. **20**(1), 83–106 (2011)

32. Tian, H., Zhang, W.: Extending l-diversity to generalize sensitive data. Data Knowl. Eng. **70**(1), 101–126 (2011)

33. Wichmann, H.-E.E., Kuhn, K.A., Waldenberger, M., Schmelcher, D., Schuffenhauer, S., Meitinger, T., Wurst, S.H., Lamla, G., Fortier, I., Burton, P.R., Peltonen, L., Perola, M., Metspalu, A., Riegman, P., Landegren, U., Taussig, M.J., Litton, J.-E.E., Fransson, M.N., Eder, J., Cambon-Thomsen, A., Bovenberg, J., Dagher, G., van Ommen, G.-J.J., Griffith, M., Yuille, M., Zatloukal, K.: Comprehensive catalog of European biobanks. Nat. Biotechnol. **29**(9), 795–797 (2011)

Author Index

Printed in the United States
By Bookmasters

Printed in the United States
By Bookmasters